Medical Informatics

Practical Guide for the
Healthcare Professional
2007

Robert Hoyt MD FACP

Melanie Sutton PhD

Ann Yoshihashi MD FACE

School of Allied Health and Life Sciences
Medical Informatics Program
University of West Florida
Pensacola Florida

Medical Informatics
Practical Guide for the Healthcare Professional

Disclaimer

The views expressed herein by the authors of this book do not necessarily reflect those of the University of West Florida.

The authors have no conflicts of interest to report. Specifically, they have no interest in, nor are they consultants for any technology companies.

Every effort has been made to make this book as accurate as possible but no warranty is implied. The information provided is on an "as is" basis. The authors and the publisher shall have neither liability nor responsibility to any person or entity with respect to any loss or damages arising from the information contained in this book.

Acknowledgements

We would like to express thanks to our families for their patience and understanding during the eighteen months of nights and weekends it took to research and write this book

Preface

We are dedicating this book to healthcare workers who are trying to improve the field of medicine through the enthusiastic, yet careful, use of technology. This book was initially written to augment our Introduction to Medical Informatics online course at the University of West Florida. We later felt that the book would also be helpful for the average healthcare worker. Many serious issues face medicine today that potentially could benefit from information technology. We hope that our book will be an important resource for all healthcare and information technology professionals and not just clinicians. This book will present a balanced scorecard of both the successes and the failures reported in the medical literature and lay press.

Our introductory book is intended to complement and not replace classic textbooks on Medical Informatics that are more technically oriented. Instead of discussing Medical Informatics theory, we will be exploring information technology applications that are cutting edge and practical. We have made this book available in a downloadable and paperback format. It is our contention that only an e-book can be current in this topic; in contrast to a standard textbook that is outdated by the time of publication. E-books also have the advantage of being updated frequently, as is our intention. Every attempt has been made to cite the most up-to-date books, lay and medical journals and web resources, resulting in more than 750 references. If readers desire more information on any topic they can access a multitude of resources in the reference sections with the electronic version hyperlinked to articles published on the Web.

We will begin with an overview of the field of Medical Informatics and culminate in the final chapter with a discussion of important emerging trends. In the middle chapters we will tackle many current and controversial issues such as electronic health records, electronic prescribing, patient safety and pay for performance. The emphasis will always be on medical issues that the average clinician or hospital faces today with solutions that are easy to understand. If we can introduce you to a new concept or a new technological tool, we will consider our work successful.

We are donating all proceeds from this book to furthering the advancement of Medical Informatics in the United States and abroad.

Table of Contents

# Chapter 1: Overview of Medical Informatics

Learning Objectives

After reading this chapter the reader should be able to:
- State the definition and origin of Medical Informatics
- Identify the forces behind Medical Informatics
- Describe the key players involved in and supporting Medical Informatics
- List the barriers to health information technology (HIT) Adoption
- Describe the educational and career opportunities in Medical Informatics

Introduction

Medical informatics has evolved as a new field in a relatively short period of time. Its emergence is partly due to the multiple problems facing the practice of medicine today. As an example, clinicians need to: be more efficient, migrate away from paper based records, reduce medication errors and have educational information and patient related data at their fingertips. Clearly, technology has the potential to help with each of those areas. With the advent of the Internet, high speed computers, voice recognition, wireless and mobile technology, clinicians and others have many more tools now available at their disposal. It would be easy to argue that information technology has advanced faster than the average healthcare worker can assimilate it into the practice of medicine. In this chapter we will present an overview of Medical Informatics, with emphasis on the factors that helped create this new field and the key players involved.

Definition

The definition of Medical Informatics is dynamic due to the rapidly changing nature of both medicine and technology. The following are three definitions frequently cited:

> "scientific field that deals with resources, devices and formalized methods for optimizing the storage, retrieval and management of biomedical information for problem solving and decision making" [1]

> "application of computers, communications and information technology and systems to all fields of medicine - medical care, medical education and medical research" [2]

"understanding, skills and tools that enable the sharing and use of information to deliver healthcare and promote health" [3]

Medical Informatics is also known as *Healthcare Informatics* and *Bioinformatics* in some circles. However, a majority agrees that the field of *Bioinformatics* involves the integration of biology and technology and can be defined as the:

"analysis of biological information using computers and statistical techniques; the science of developing and utilizing computer databases and algorithms to accelerate and enhance biological research" [4]

Some prefer the designation of *Biomedical Informatics* because it encompasses both Medical Informatics and Bioinformatics. Under this more global term are dental informatics, nursing informatics, public health informatics, pharmacy informatics, medical imaging informatics, veterinary informatics, proteomics and genomics.[5] As we move closer to integrating human genetics into the day-to-day practice of medicine this more global definition may gain traction. We have chosen to use Medical Informatics throughout the book for consistency.

The reality is that Medical Informatics is all about providing medical information to healthcare workers in their workspace, anytime, anywhere to assist in education, productivity, patient safety, quality of care and research. Medical Informatics emphasizes bi-directional *information brokerage* to assist all healthcare workers.

Background

Given the fact that most businesses incorporate technology into their enterprise fabric, one could argue that it was just a matter of time before the tectonic forces of medicine and technology collide. One of the reasons Medical Informatics is in the spotlight is the fact that more healthcare workers will need to be "bilingual" in both technology and medicine. With the increasing use of technology, more skilled healthcare workers will be needed for implementation and training. Vendors, insurance companies and governmental organizations will also be looking for the same expertise. In 2005 the American Medical Informatics Association created the *AMIA 10x10 Program* with the goal of training 10,000 healthcare workers in Medical Informatics over the next 10 years. [6] This timeline corresponds to the national plan to have a universal interoperable electronic health record system by 2014.[7]

Information Technology (IT) interfaces with and connects many important functions in healthcare organizations (figure 1.1). It should be pointed out however that IT is not the center of the organization, just a common thread.

This is undoubtedly the reason why The Joint Commission created a Management of Information standard for hospital certification. [8]

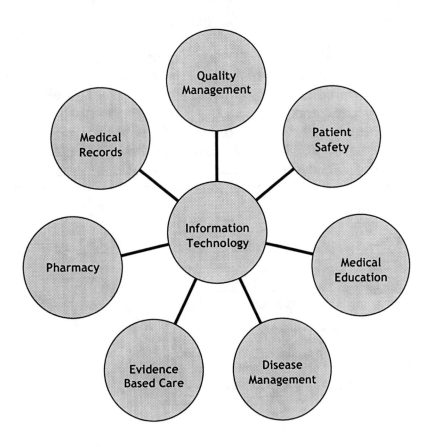

Figure 1.1. Healthcare functions and information technology

There are multiple forces driving Medical Informatics, but the major ones are the desire to decrease costs, improve patient safety and improve the quality of medical care. In this book we will discuss each driving force and their inter-relationships. In addition to these three forces, the natural diffusion of technology also exerts an influence. As technologies like wireless and voice recognition improve and mature, new applications will arise. Vendors look for new niches to introduce and promote these new innovations. Although Moore's Law relates to transistors in computer chips it tends to hold true of many aspects of technology.

> " the number of transistors on a computer chip doubles every 1.5 years. He also observed that the number of instructions per second performed by a chip also doubles every 1.5 to 2 years" [9]

Technology will continue to evolve at a rapid rate but it is important to realize that it often advances in an asynchronous manner. For example, tablet and laptop personal computers (PCs) have come a long way with excellent processor speed and memory but their utility is limited due to a battery life of only several hours. Tablet PCs would be appealing to nurses but most nurses now work 8+ hour shifts. When improved battery life becomes a reality it will allow for more technologies to be maximally utilized.

The electronic health record (EHR), covered in chapter two, could be considered the centerpiece of Medical Informatics with its potential to improve patient safety, productivity and data retrieval. The EHR will likely be the focal point of all patient encounters in the future (Figure 1.2). Multiple resources that are currently standalone programs will be incorporated into the EHR. It is anticipated that the EHR will eventually be shown to improve patient outcomes. Not included in the diagram is genetic information that will one day become routine medical information and part of everyone's medical record.

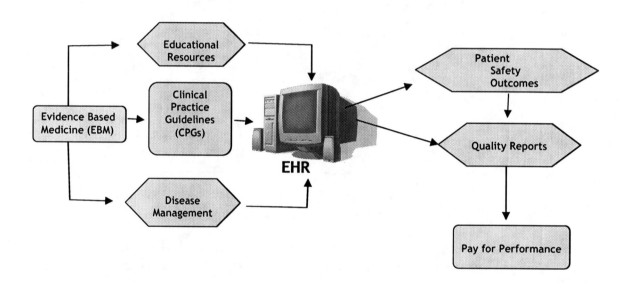

Figure 1.2. Electronic health record (EHR) dynamics.

It is also important to realize that one of the outcomes of widespread use of EHRs will be the production of voluminous healthcare data. As pointed out by Steve Balmer, the CEO of Microsoft, there will be an "explosion of data" as a result of automating multiple medical processes.[10] This explains, in part, why technology giants such as Microsoft, IBM and Google are now entering the healthcare arena. Vendors are beginning to offer comprehensive medical software programs for free in exchange for mining de-identified patient data.

Historical Highlights

Information technology has been pervasive in the field of Medicine for only about two decades. During this time we have experienced astronomical advances in technology to include personal computers, high resolution imaging, the Internet and wireless, to mention a few. Early on there was no unifying plan or vision as to how to advance healthcare delivery using information technology. Nevertheless, forward thinking scientists created the technology we take for granted today. The following are some of the more noteworthy developments in Medical Informatics and information technology:

- The first general purpose computer (ENIAC) was released in 1946 and required 1000 sq ft of floor space. IBM developed the first personal computer in 1982 with a total of 16 K of memory. [11] Computers were first theorized to be useful for medical diagnosis and treatment by Ledley and Lusted in the 1950's.[12] They reasoned that computers could archive and process information more rapidly than humans

- It is thought that the origin of the term Medical Informatics dates back to the 1960's in France ("Informatique Medicale")[13]

- In the mid-1960's MEDLINE and MEDLARS were created to organize the world's medical literature. For older clinicians who can recall trying to research a topic using the multi-volume text *Index Medicus*, this represented a quantum leap forward

- Artificial intelligence medical projects such as MYCIN (University of Pittsburg) and INTERNIST-1 (Stanford) appeared in the 1970's [14]

- The development of the Internet began in 1969 with the creation of the government project ARPANET.[15] The World Wide Web was conceived by Tim Berners-Lee in 1990 and the first web browser Mosaic appeared in 1993.[16-17] The Internet became the backbone for digital libraries used by clinicians and patients

- The electronic health record has been discussed since the 1970's and recommended by the Institute of Medicine in 1991.[18] EHRs will be discussed in detail in chapter two

- The PalmPilot PDA appeared in 1996 as the first truly popular handheld computing device.[19] PDAs loaded with medical software are now standard equipment for any resident in training. They will be discussed in more detail in a later chapter.
- In 2003 the Human Genome Project (HGP) was completed after thirteen years of international collaborative research. To map all human genes was one of the greatest accomplishments in scientific history. Although the project is complete, it will take years to analyze the data.[20] Many experts feel that the HGP will have a major impact on how we practice medicine. The HGP will be discussed in the chapter on Bioinformatics

- The concept of the National Health Information Infrastructure (renamed the National Health Information Network or NHIN) appeared in 2004. The goal of the NHIN is to connect all electronic health records and regional networks together into one system, in one decade.[21] Achieving interoperability among all healthcare systems and workers in the United States would be of monumental significance. This will be discussed in more detail in several other chapters

Key Players

Information Technology is important to multiple players in the field of Medicine. The goals of these different groups are to:

- reduce medical errors and resultant litigation
- provide better return on investment (ROI)
- improve communication among the key players
- improve the quality of care
- reduce duplication of tests or prescriptions ordered
- improve patient outcomes
- standardize care among clinicians, organizations and regions
- speed up access to care and administrative transactions
- protect privacy and ensure security

Next is a list of the key players and how they utilize information technology (adapted from *Crossing the Quality Chasm*)[22]

Patients
- Online searches for health information
- Web portals for storing personal medical information, making appointments, checking lab results, e-visits, etc
- Research choice of physician, hospital or insurance plan
- Online patient surveys
- Online chat, blogs, podcasts and support groups
- Personal health records
- Telemedicine and home Telemonitoring

Clinicians
- Medline searches
- Online resources and digital libraries
- Patient web portals, secure e-mail and e-visits
- Physician web portals
- Clinical decision support, e.g. reminders and alerts
- Electronic health records (EHRs)
- Personal Digital Assistants (PDAs) with medical software

- Telemedicine
- Online continuing medical education (CME)
- Electronic (e)-prescribing
- Disease management and registries
- Picture archiving and communication systems (PACS)
- Pay for performance
- E-research

Support Staff
- Patient enrollment
- Electronic appointments
- Billing
- EHR
- Credentialing
- Inputting vital signs
- Practice management software
- Secure patient-office e-mail communication

Public Health
- Incident reports
- Syndromic surveillance as part of bio-terrorism program
- Establish link to all public health departments (National Public Health Information Network)
- Centers for Disease Control and Prevention (CDC) reports
  - Morbidity and Mortality Weekly Report (MMWR) reports

Medical Education
- Online medical resources for clinicians, patients and staff
- Online CME
- Medline searches
- Video teleconferencing, web conferencing, podcasts, etc

The Joint Commission
- Promotes the need for information systems to monitor quality and patient safety and framework for "pay for performance"
- Promotes information systems to also decrease liability
- Seeks improved medication safety through technology
- Monitors the management of information in hospitals

Insurance Companies
- Electronic claims transmission
- Trend analysis
- Physician profiling
- Information systems for "pay for performance"

- Monitor adherence to clinical guidelines
- Monitor adherence to preferred formularies
- Promote claims based personal health records
- Monitor use of computer order entry
- Reduce litigation by improved patient safety through fewer medication errors

State and Federal Governments
- Promote universal and interoperable electronic health records
- Create a national healthcare information network (NHIN)
- Reduce preventable deaths with better clinical decision support
- Reduce cost by reducing redundancy and litigation and promoting better adherence to automated clinical practice guidelines
- Reward quality through "pay for performance"
- Promote e-prescribing

Hospitals
- Interoperable electronic health records
- Electronic billing
- Information systems to monitor outcomes, length of stay, disease management, etc
- Bar coding and radio frequency identification (RFID) to track patients, medications, assets, etc
- Wireless technology
- Technology to reduce errors and litigation
- Technology to reduce cost
- Patient and physician portals
- E-prescribing
- Part of Health Information Exchanges (HIEs)
- Telemedicine
- Picture archiving and communication systems

Research
- Database creation to study populations, genetics and disease states
- Online collaborative web sites e.g. Microsoft SharePoint
- Electronic forms e.g. Microsoft InfoPath
- Software for statistical analysis of data e.g. SPSS
- Literature searches
- Improved subject recruitment using EHRs and e-mail
- Online submission of grants

Technology Vendors
- Applying new technology innovations in the field of medicine; hardware, software, genomics, etc

- Data mining
- Interoperability

Academic, Private and Governmental Influences

Institute of Medicine (IOM). It could be easily argued that the organization in the United States most responsible for promoting the use of information technology in the field of medicine is the Institute of Medicine. It was established in 1970 by the National Academy of Sciences with the task of evaluating policy relevant to healthcare and providing feedback to the Federal Government. In their two pioneering books *To Error is Human* (1999) and *Crossing the Quality Chasm* (2001), they reported that approximately 98,000 deaths occur yearly due to medical errors. It is their contention that an information technology infrastructure will help the six aims set forth by the IOM: safe, effective, patient centered, timely, efficient and equitable medical care. The infrastructure would support "efforts to re-engineer care processes, manage the burgeoning clinical knowledge base, coordinate patient care across clinicians and settings over time, support multidisciplinary team functioning and facilitate performance and outcome measurements for improvement and accountability". They also stress "the importance of building such an infrastructure to support evidence based practice, including the provision of more organized and reliable information sources on the Internet for both consumers and clinicians and the development and application of decision support tools". [22, 24]

Two of the IOM's twelve executive recommendations directly relate to information technology:

<u>Recommendation 7:</u> "improve access to clinical information and support clinical decision making"

<u>Recommendation 9:</u> "Congress, the executive branch, leaders of health care organizations, public and private purchasers and health informatics associations and vendors should make a renewed national commitment to building an information infrastructure to support health care delivery, consumer health, quality measurement and improvement, public accountability, clinical and health services research, and clinical education. This commitment should lead to *the elimination of most handwritten clinical data by the end of the decade*" [22]

The IOM cites twelve information technology applications that might narrow the quality chasm. Many of these will be discussed in other chapters:

1. Web based "Personal Health Records"
2. Patient access to hospital information systems to access their lab and x-ray reports
3. Access to general health information via the Internet
4. Electronic medical record with clinical decision support

5. Pre-visit online histories
6. Inter-hospital data sharing, e.g. lab results
7. Information to manage populations with patient registries and reminders
8. Patient-physician electronic messaging
9. Online data entry by patients for monitoring, e.g. glucose results
10. Online scheduling
11. Computer assisted telephone triage and assistance (nurse call centers)
12. Online access to clinician or hospital performance data [23]

The Association of American Medical Colleges (AAMC). For more than twenty years the AAMC has been an advocate of incorporating informatics into medical school curricula and promoting medical informatics in general. In their *Better Health 2010 Report* they make the following recommendations:
- Optimize the health and healthcare of individuals and populations through best practice information management
- Enable continuous and life-long performance based learning
- Create tools and resources to support discovery, innovation and dissemination of research results
- Build and operate a robust information environment that simultaneously enables healthcare, fosters learning and advances science [25]

Bridges to Excellence. This organization consists of employers, physicians, health plans and patients. They currently have three programs with incentives for technology adoption, improved diabetic and cardiac care. Their vision is as follows:
- "Reengineering care processes to reduce mistakes will require investments, for which purchasers should create incentives
- Significant reductions in defects (misuse, under use, overuse) will reduce the waste and inefficiencies in the health care system today
- Increased accountability and quality improvements will be encouraged by the release of comparative provider performance data, delivered to consumers in a compelling way" [26]

eHealth Initiative. This is a non-profit organization promoting the use of information technology to improve quality and patient safety. Its membership includes virtually all stakeholders involved in the delivery of healthcare. This organization created the "Connecting Communities for Better Health Program" that provides seed money to support and connect disparate healthcare communities. [27]

Leapfrog. Leapfrog is a consortium of over one hundred and seventy major employers seeking to purchase the highest quality and safest healthcare. Voluntary reporting by hospitals has made hospital comparisons possible and the results are reported on their website. They also have a hospital rewards program to provide incentives to hospitals that show they deliver quality care.

One of their patient safety measures is the use of computerized physician order entry (CPOE) by hospitals that will be covered in several other chapters.[28]

US Federal Government. In July 2004 Dr David Brailer was appointed by President Bush as the first National Coordinator for Health Information Technology (NCHIT) under the Department of Health and Human Services (HHS). The most significant goal of the Office of the National Coordinator for Health Information Technology (ONC) [29] is the creation of a universal interoperable electronic health record by the year 2014. Dr Brailer promoted many decentralized IT initiatives during his 2 year tenure. Dr. Robert Kolodner, former chief health informatics officer at Veterans Health Affairs, was appointed his permanent successor in April 2007 after serving as interim NCHIT since September 2006. The ONC published the *Framework for Strategic Action* (July 2004) with the following goals:

1. *Inform Clinical Practice*
   - Incentivize EHR adoption
   - Reduce risk of EHR investment
   - Promote EHR diffusion in rural and underserved areas
2. *Interconnect Clinicians*
   - Foster regional collaborations
   - Develop a national health information network
   - Coordinate federal health information systems
3. *Personalize Care*
   - Encourage use of personal health records
   - Enhance informed consumer choice
   - Promote use of telehealth systems
4. *Improve Population Health*
   - Unify public health surveillance architectures
   - Streamline quality and health status monitoring
   - Accelerate research and dissemination of evidence

A useful list of all federal health information technology (HIT) web links can be found on the ONCHIT website.[30]

Agency for Healthcare Research and Quality (AHRQ). The AHRQ is an agency under the Department of Health and Human Services. It is "the lead Federal agency charged with improving the quality, safety, efficiency, and effectiveness of health care for all Americans. As one of 12 agencies within the Department of Health and Human Services, AHRQ supports health services research that will improve the quality of health care and promote evidence based decision making". This agency set aside $139 million in grant money in 2005 to support healthcare information technology (HIT) over the next five years. AHRQ also maintains the National Resource Center for HIT and an extensive patient safety and quality section.[31]

Centers for Medicare and Medicaid Services (CMS). The CMS also falls under the Department of Health and Human Services. In an effort to improve quality and decrease costs, CMS has several pilot projects to include new "pay for performance" demonstration projects that link payments to improved patient outcomes. This will be discussed in more detail in the chapter on pay for performance.[32]

Commission on Systemic Interoperability. This is a commission of 14 members tasked to formulate the strategy for national healthcare information technology.[33] They are charged with:
- Creating the strategy for an information network
- Determining the costs and benefits of the network
- Evaluating the barriers and opportunities of creating the network

The commission reported fourteen recommendations about adoption, interoperability and connectivity on October 25th 2005.[34-35] This commission will be discussed in greater detail in the chapter on interoperability.

The American Health Information Community (AHIC). Secretary Mike Leavitt of the Department of Health and Human Services in September 2005 created the AHIC. It is a seventeen member advisory board tasked to advance the United State's health IT agenda. Representatives are from government agencies, patient advocacy groups, private employers, insurance companies and the technology industry.[36-37]

Health IT Legislation. The US Congress has been fully supportive of healthcare information technology. What is noteworthy is not only is there evidence of interest in healthcare informatics by Congress; most of the proposed legislation is bi-partisan. While numerous pieces of legislation have been introduced, few have been written into law, however.

Barriers to Health Information Technology (HIT) Adoption

According to Anderson, the United States is at least 12 years behind many industrialized nations in terms of HIT adoption. Total investment in 2005 per capita was 43 cents compared to $21 for Canada, $4.93 for Australia, $21 for Germany and $192 for the United Kingdom.[38] Healthcare information technology adoption has multiple barriers listed below and discussed in later chapters:
- **Inadequate time.** This complaint is a common thread that runs throughout most discussions of technology barriers. Busy clinicians complain that they don't have enough time to read, learn new technologies or research vendors. They are also not reimbursed to become technology experts

- **Cost.** It is estimated that a national health information network (NHIN) will cost $156 billion dollars over five years and $48 billion annually in operating expenses. [39] Technologies such as picture archiving and communications systems (PACS) and electronic health records also suffer from high price tags

- **Lack of interoperability.** A true national healthcare information network cannot occur until data standards are adopted and implemented nation wide. This is covered in more detail in chapter three

- **Change in workflow.** Significant changes in workflow will be required to integrate technology into the inpatient and outpatient setting. As an example, clinicians may be accustomed to ordering lab or x-rays by giving a handwritten request to a nurse. Now they have to learn to use computerized physician order entry (CPOE). According to Dr Carolyn Clancy, the director of AHRQ:

  "The main challenges are not technical; it's more about integrating HIT with workflow, making it work for patients and clinicians who don't necessarily think like the computer guys do" [40]

- **Privacy.** The Health Information Portability and Accountability Act (HIPAA) of 1996 was designed, in part, for the electronic transmission of personal health information. This Act has caused healthcare organizations to re-think every issue related to information privacy and security. This will be covered in more detail in chapter three

- **Legal.** The Stark and Anti-kickback laws prevent hospital systems from providing or sharing technology such as computers and software with referring physicians. [41] Exceptions have been made to these laws in 2006, as will be pointed out in other chapters. This is particularly important for hospitals in order to share electronic health records and e-prescribing

- **Behavioral change.** Perhaps the most challenging barrier is behavior. Dr. Frederick Knoll of Stanford University described the five stages of medical technology acceptance: (1) abject horror, (2) swift denunciation, (3) profound skepticism, (4) clinical evaluation, then, finally (5) acceptance as the standard of care. [42] It is unrealistic to expect all medical personnel to embrace technology. In 1962 Everett Rogers wrote *Diffusion of Innovations* in which he delineated different categories of acceptance of innovation:
    o the innovators (2.5%) are so motivated they may need to be slowed down
    o early adopters (13.5%) accept the new change and teach others

- o early majority adopters (34%) require some motivation and information from others in order to adopt
- o the late majority (34%) require encouragement to get them to eventually accept the innovation
- o laggards (16%) require removal of all barriers and often require a direct order [43]

It is important to realize, therefore, that at least 50% of medical personnel will be slow to accept any information technology innovations and they will be perceived as dragging their feet. This is where selecting clinical champions and intensive training play a major role in implementation.

- **Inadequate workforce.** As pointed out by Dr. William Hersh of the Oregon Health and Science University, there is a need for a work force capable of leading implementation of the electronic health record and other technologies.[44] The first Work Force for Health Information Transformation Strategy Summit, hosted by the American Medical Informatics Association (AMIA) and the American Health Information Management Association (AHIMA) made several strategic recommendations regarding how to improve the work force.[45] The American Medical Informatics Association has developed the *AMIA 10x10 Program* with the intent of developing ten thousand knowledgeable healthcare workers by the year 2010 [46]

- **Over-hyped technology.** This is not listed as a true barrier but more a note of caution. The Gartner Group describes 5 phases of the hype cycle that outline the progression of technology from excitement to disillusionment to final reality. [47] Physicians tend to be leery about new technologies that promise a lot, but may deliver little. As a rule, if it doesn't save time or money they are not interested. Be wary of vendors who promise the world and medical experts who are also consultants for IT companies

Medical Informatics Programs

One of the best sites to review the various Medical Informatics programs in the United States and overseas can be found on the American Medical Informatics Association's web site (www.amia.org). [48] Another excellent site for listing available Medical Informatics programs in the United States and the United Kingdom is the www.Biohealthmatics.com web site.[49] Medical Informatics programs can be degree, certificate, fellowship and short courses. Most programs are part of a medical or nursing school and others may be part of a health related organization such as the National Library of Medicine. Courses can be online or taught in a classroom setting. Medical Informatics degree programs are available as follows: associate degree, undergraduate degree,

Master's degree, PhD degree or as part of another degree program. The following table will give you an idea of how many programs are available in North America and in which category. [48]

Program type	Number of programs
Associate degree	1
Undergraduate degree	3
Masters degree	54
PhD degree	25
Certificate	31
Short courses	13
Online courses	21

Table 1.1. Medical Informatics programs listed on the AMIA web site (August 2006)

Medical Informatics Organizations

The following organizations are considered among the most important and influential in health information technology

American Medical Informatics Association (AMIA)
- Founded in 1990
- As of 2006 has greater than 3500 members
- Members are from 42 countries
- Web site includes job exchange, academic programs, fellowships, grants, and an e-newsletter
- Membership includes subscription to the Journal of the American Medical Informatics Association (JAMIA)
- Opportunity to join a working group to discuss issues and formulate white papers
- Annual national symposium in the fall as well as a Spring Congress [48]

Healthcare Information and Management Systems Society (HIMSS)
- Founded in 1961
- As of 2006 has over 17,000 members
- 275 corporations are members
- Annual symposium with more than 20,000 attendees
- Professional certification
- Educational publications, books and CD-ROMs
- Web conferences on Medical Informatics topics
- Surveys on multiple topics [50]

American Health Information Management Association (AHIMA)
- As of 2006 has more than 50,000 members

- It began in 1928 as a medical records association. It now includes any healthcare worker involved in information management
- "AHIMA supports the common goal of applying modern technology to and advancing best practices in health information management"
- Has an excellent section on personal health records [45]

Alliance for Nursing Informatics (ANI)
- Combines 20 separate nursing informatics organizations
- As of 2006 has more than 3,000 members
- Sponsored by both the AMIA and HIMSS
- Provides a collaborative group for consensus about nursing informatics [51]

Careers

The timing is excellent for a career in Medical Informatics. With the emphasis on a national healthcare information network and electronic medical records, hospitals and vendors will be looking for healthcare workers who are knowledgeable in both technology and medicine. The Biohealthmatics, HIMSS, American Nurse Informatics and the AMIA web sites list multiple interesting healthcare IT jobs. Examples include nurse and physician informaticists, systems analysts, information directors, chief information officers (CIOs) and chief medical information officers (CMIOs). [48-51] Recruiting organizations also maintain multiple listings for healthcare IT jobs.

The American Medical Informatics Association is in the process of establishing the medical subspecialty of clinical informatics. It is likely that it will take several years to have this new specialty approved by the American Board of Medical Specialties. [52] Although physicians can become chief medical information officers in very large organizations, the reality is that nurses have the greatest potential to be involved with IT training and implementation at the average hospital or large clinic. Of note is the fact that the field of nursing already has an informatics specialty certification. IT Nurses today are likely to help implement IT initiatives such as EHRs, bar coding and e-prescribing. We believe that nurses are best positioned to assist with future IT initiatives and have the greatest need for formal training. A 2007 HIMSS Nursing Informatics Survey (776 respondents) revealed some interesting statistics:
- 33% of today's nurse informaticists have 10 years in the specialty
- 57% involved with EHRs
- 54% work in a hospital setting
- 48% of today's nurse informaticists have at least 16 years of clinical nursing experience
- 34% have formal IT training; most receive on the job training
- 74% do not continue clinical duties after the transition to their IT position
- The average salary was $83,675 [53]

Resources

We have tried to create a helpful online resource web site to augment this book. Valuable web links are organized in a similar manner as the book chapters noted in the Table of Contents. We have also added links to excellent newsletters and journals. It is extremely helpful to receive e-newsletters that update national Medical Informatics initiatives on a regular basis. Standard print textbooks are not listed, because in our opinion, they are outdated too rapidly. Moreover, many of the recent and important changes in Medical Informatics cannot be found in the medical literature. The resource site is hosted on the University of West Florida's Medical Informatics web site www.uwf.edu/sahls/medicalinformatics/ [54]

Conclusion

Medical Informatics is a new, exciting and evolving field. In spite of its importance and popularity, significant obstacles remain. The expectation is that information technology will improve medical quality, patient safety, educational resources and patient-physician communication while decreasing cost. Although technology holds great promise, it is not the solution for every problem facing medicine today. As noted by Dr. Safron of the American Medical Informatics Association "technology is not the destination, it is the transportation".[48] Research in Medical Informatics is being published at an increasing rate so that new technologies are being evaluated more objectively. Better studies are needed to demonstrate the effects on actual patient outcomes and return on investment. The available evidence is frequently flawed because information is based on surveys or expert opinion with inherent shortcomings.

References

1. Shortliffe, E What is medical informatics? Lecture. Stanford University,1995.
2. MF Collen, Preliminary announcement for the *Third World Conference on Medical Informatics, MEDINFO 80*, Tokyo
3. UK Health Informatics Society http://www.bmis.org (Accessed September 5 2005)
4. Center for Toxicogenomics http://www.niehs.nih.gov/nct/glossary.htm (Accessed September 10 2005)
5. Biohealthmatics http://www.biohealthmatics.com/knowcenter.aspx (Accessed September 5 2006)
6. American Medical Informatics Association http://www.amia.org (Accessed September 15 2006)
7. Dept. Health and Human Services http://www.hhs.gov/news/press/2005pres/20050603.html (Accessed September 8 2005)
8. The Joint Commission http://www.jcrinc.com/8250/ (Accessed March 18 2007)
9. Intel http://www.intel.com/technology/silicon/sp/glossary.htm (Accessed September 4 2005)
10. Balmer S Keynote Address 2007 HIMSS Conference. February 26 2007

11. A history of computers http://www.maxmon.com/history.htm (Accessed September 30 2005)

12. Hersch WR Informatics: Development and Evaluation of Information Technology in Medicine JAMA 1992;267:167-70

13. VUMC Dept. of Biomedical Informatics http://www.mc.vanderbilt.edu/dbmi/informatics.html (Accessed Oct 1 2005)

14. Health Informatics http://en.wikipedia.org/wiki/Medical_informatics (Accessed September 20 2005)

15. Howe, W. A Brief History of the Internet http://www.walthowe.com/navnet/history.html (Accessed September 24 2005)

16. Zakon, R Hobbe's Internet Timeline v8.1 http://www.zakon.org/robert/internet/timeline (Accessed September 24 2005)

17. W3C http://www.w3.org/WWW/ (Accessed September 25 2005)

18. Advance for Healthcare Executives http://www.xwave.com/healthcare/cms/about_us/doc/industry_analysis_electronic_medical_record_system_3.doc (Accessed October 1 2005)

19. Koblentz, E The Evolution of the PDA http://www.snarc.net/pda/pda-treatise.htm (Accessed Oct 3 2005)

20. Human Genome Project. US Dept of Energy http://www.ornl.gov/sci/techresources/Human_Genome/home.shtml (Accessed Oct 5 2005).

21. FAQ's about NHII. Dept. of Health and Human Services. http://aspe.hhs.gov/sp/nhii/FAQ.html (Accessed September 28 2005)

22. Crossing the Quality Chasm: A new health system for the 21th century (2001) The National Academies Press http://www.nap.edu/books/0309072808/html/ (Accessed October 5 2005)

23. Crossing the Chasm with Information Technology. Bridging the gap in healthcare. First Consulting Group July 2002 http://www.chcf.org/documents/ihealth/CrossingChasmIT.pdf (Accessed September 20 2005)

24. To Error is Human: Building a safer Healthcare System (1999) The National Academies Press http://www.nap.edu/catalog/9728.html (Accessed October 5 2005)

25. Association of American Medical Colleges. Better health 2010. http://www.aamc.org/programs/betterhealth/betterhealthbook.pdf (Accessed October 4 2005)

26. Bridges To Excellence http://www.bridgestoexcellence.org/bte/bte_overview.htm (Accessed October 10 2005)

27. E-health Initiative http://www.ehealthinitiative.org/default.mspx (Accessed October 10 2005)

28. The Leapfrog Group http://www.leapfroggroup.org/ (Accessed October 5 2005)

29. Office of the National Coordinator for Health Information Technology http://www.hhs.gov/healthit/ (Accessed October 12 2005)

30. ONCHIT. Directory of Federal Health IT Programs http://www.hhs.gov/healthit/federalprojectlist.html#intitiativestable (Accessed October 20 2005)

31. Agency for Healthcare Research and Quality http://www.ahrq.gov/ (Accessed October 12 2006)

32. Centers for Medicare and Medicaid Services. Medicare Demonstrations. http://www.cms.hhs.gov/researchers/demos/mma646/ (Accessed October 13 2005)

33. National Library of Medicine. Commission on Systematic Interoperability Homepage. http://www.nlm.nih.gov/csi/csi_home.html (Accessed October 12 2005)

34. Commission on Systematic Interoperability. Ending the Document Game http://endingthedocumentgame.gov/ (Accessed November 1 2005)

35. Weier S. Commission releases 14 Interoperability Recommendations. October 26 2005) http://www.ihealthbeat.org/index.cfm?Action=itemPrint&itemID=115499 (Accessed October 27 2005)

36. ONCHIT. American Health Information Community http://www.hhs.gov/healthit/ahic.html (Accessed November 2 2005)

37. American Health Information Community Charter.
http://www.hhs.gov/healthit/ahiccharter.pdf (Accessed November 2 2005)
38. Anderson GF et al. Health Care Spending and use of Information Technology in OECD
countries. Health Affairs 2006;25:819-831
39. Basch P et al .Electronic health records and the national health information network:
affordable, adaptable and ready for prime time? Ann Intern Med 2005 143(3):165-73
40. Interview with Dr. Carolyn Clancy. Medscape June 2005. www.medscape.com (Accessed
November 4 2005)
41. Physician's News Digest http://www.physiciansnews.com/law/1200maruca.html (Accessed
November 3 2005)
42. Medical Imaging in the Age of Informatics
http://projects.ics.hawaii.edu/strev/ics691/presentations/ (Accessed November 15 2005)
43. Rogers EM, Shoemaker FF Communication of Innovation 1971 New York, The Free Press
44. Hersh W .Health Care Information Technology JAMA 2004; 292 (18):2273-441
45. American Health Information Management Association site
http://www.ahima.org (Accessed November 14 2006)
46. AMIA 10x10 Program. http://www.amia.org/10x10/partner.asp (Accessed August 14 2006)
47. Gartner hype cycle http://gsb.haifa.ac.il/~sheizaf/ecommerce/GartnerHypeCycle.html
(Accessed November 21 2005)
48. AMIA www.amia.org (Accessed February 10 2007)
49. Biohealthmatics www.Biohealthmatics.com (Accessed November 15 2006)
50. Health Information Management Systems Society www.himss.org (Accessed November 16
2006)
51. Alliance for Nursing Informatics http://www.allianceni.org/ (Accessed November 16 2006)
52. Clinical Informatics Specialty on Drawing Board. April 3 2007 www.ihealthbeat.org
(Accessed April 6 2007)
53. 2007 HIMSS Nursing Informatics Survey
http://www.himss.org/content/files/surveyresults/2007NursingInformatics.pdf (Accessed
March 3 2007)
54. University of West Florida. Introduction to Medical Informatics's resource page
www.uwf.edu/sahls/medicalinformatics/ (Accessed March 1 2007)

# Chapter 2: Electronic Health Records

Learning Objectives

After reading this chapter the reader should be able to:
- State the definition and history of electronic health records
- Describe the limitations of paper based health records
- Identify the benefits of electronic health records
- List the key components of an electronic health record
- Describe the benefits of computerized order entry and clinical decision support systems
- State the obstacles to purchasing and implementing an electronic health record

Introduction

> "If computers get too powerful, we can organize them into a committee. That will do them in"
>
> Bradley's Bromide

There is no topic in Medical Informatics as important, yet controversial as the electronic health record (EHR). That is the reason EHRs will be discussed in the second chapter. In 1970 Schwartz predicted "clinical computing would be common in the not too distant future".[1] In 1991 the Institute of Medicine (IOM) recommended electronic health records as a solution for many of the problems facing modern medicine.[2] In spite of this IOM recommendation, little progress has been made during the last decade for multiple reasons. As Dr. Donald Simborg states, the slow acceptance of electronic health records is like the "wave that never breaks".[3] Most recently, following hurricane Katrina in 2005, articles have appeared in the lay press promoting the importance of electronic health records during a natural disaster.[4]

The true adoption rate of ambulatory EHRs is probably 10-20% depending on which study you read and what group is studied.[5] Many of the commonly quoted statistics come from surveys with their obvious shortcomings. It is also important to realize that many outpatient practices may have EHRs but continue to run dual paper and electronic systems. It should also be noted that EHRs are being purchased largely by primary care practices as opposed to surgical specialties which may skew the statistics. Furthermore, a significant concern is that small and/or rural practices are more likely to lack the finances and information technology support to purchase and implement EHRs.

One can only speculate why the medical profession has been willing to tolerate the lack of legible and accessible information for so many years. Many physicians believe that it is not their problem and that someone else should pick up the tab. Others are concerned that they will purchase the wrong system and waste money and others are simply overwhelmed with the task of implementing and training for a completely different system. As a group, physicians are not noted for embracing innovation. In their defense, new technologies should be shown to improve patient care, save time or money in order to be accepted.

There are over two hundred EHR vendors but only about ten seem to be consistently successful in terms of a large client base. If the selection and purchase of EHRs was easy they would already be universal. As you will see later in this chapter, there are issues such as implementation and training that are just as important as the decision which EHR to purchase.

The United States is not the only country to face the challenge of trying to have a nationwide interoperable electronic health record. Canada plans a universal EHR by 2009, Australia by 2010 and Great Britain by 2014, although delays seem inevitable.[6] According to a 2000 *HarrisInteractive* study, Sweden, Germany, Great Britain and Denmark are considerably ahead of the United States in terms of general practitioners using electronic health records.[7]

There is no universally accepted definition of an EHR. As more functionality is added the definition will need to be broadened. Importantly, EHRs are also known as electronic medical records (EMRs), computerized medical records (CMRs), electronic clinical information systems (ECIS) and computerized patient records (CPRs). Throughout this book we will use electronic health record as the more accepted and inclusive term. Figure 2.1 demonstrates the relationship between the EHR, EMR and personal health record (PHR).[8]

- EHR is the larger system that includes the EMR and PHR and interfaces with multiple other electronic systems locally, regionally or nationally
- EMR is the electronic patient record located in an office or hospital
- PHR is a collection of health information by and for the patient that can be part of the EMR.

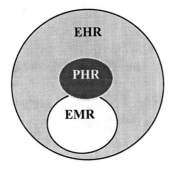

Figure 2.1. Relationship between EHR (electronic health record), PHR (personal health record) and EMR (electronic medical record)

Why do we need an electronic health record?

The following are the most significant reasons why our healthcare system would benefit from the widespread transition from paper to electronic health records.

1. The paper record is severely limited. Much of what can be said about handwritten prescriptions can also be said about handwritten office notes. Figure 2.2 illustrates the problems with a paper record. The handwriting is illegible and cannot be electronically shared or stored. Electronic patient encounters represent a quantum leap in legibility and retrievability.

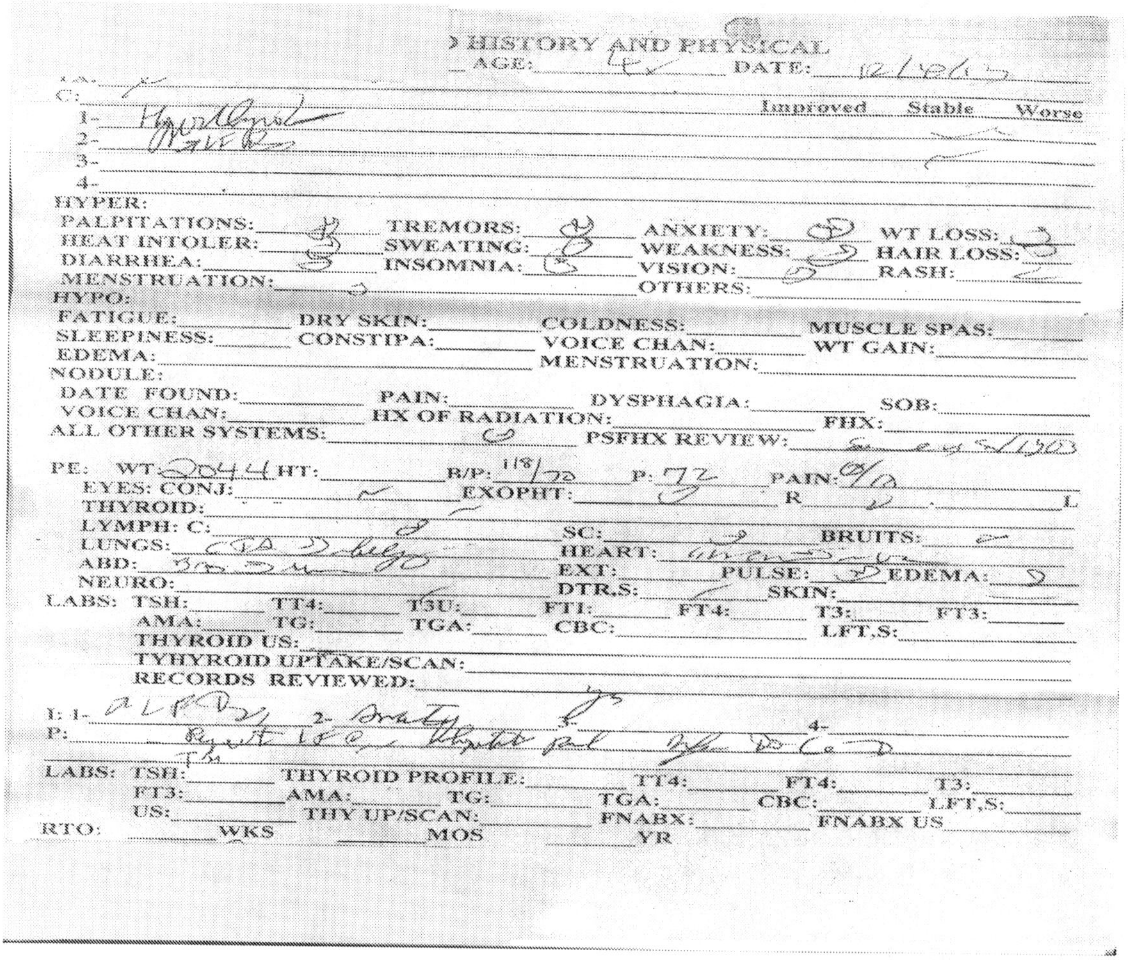
HISTORY AND PHYSICAL
AGE: DATE:
C:
Improved Stable Worse
1-
2-
3-
4-
HYPER:
PALPITATIONS: TREMORS: ANXIETY: WT LOSS:
HEAT INTOLER: SWEATING: WEAKNESS: HAIR LOSS:
DIARRHEA: INSOMNIA: VISION: RASH:
MENSTRUATION: OTHERS:
HYPO:
FATIGUE: DRY SKIN: COLDNESS: MUSCLE SPAS:
SLEEPINESS: CONSTIPA: VOICE CHAN: WT GAIN:
EDEMA: MENSTRUATION:
NODULE:
DATE FOUND: PAIN: DYSPHAGIA: SOB:
VOICE CHAN: HX OF RADIATION: FHX:
ALL OTHER SYSTEMS: PSFHX REVIEW:
PE: WT: HT: B/P: P: PAIN:
EYES: CONJ: EXOPHT: R L
THYROID:
LYMPH: C: SC: BRUITS:
LUNGS: HEART:
ABD: EXT: PULSE: EDEMA:
NEURO: DTR,S: SKIN:
LABS: TSH: TT4: T3U: FTI: FT4: T3: FT3:
AMA: TG: TGA: CBC: LFT,S:
THYROID US:
TYHYROID UPTAKE/SCAN:
RECORDS REVIEWED:
I: 1- 2- 3- 4-
P:
LABS: TSH: THYROID PROFILE: TT4: FT4: T3:
FT3: AMA: TG: TGA: CBC: LFT,S:
US: THY UP/SCAN: FNABX: FNABX US
RTO: WKS MOS YR

Figure 2.2. Outpatient paper encounter form

Almost every industry is now computerized for rapid data retrieval and trend analysis. Look at the stock market or companies like Walmart or Federal Express. Why not medicine? With the relatively recent advent of "pay for performance" there is a new reason to embrace technology in order to receive more reimbursement. It is much easier to retrieve and track patient data using EHRs and patient registries than to use labor intensive paper chart reviews. EHRs are much better organized than paper charts, allowing for faster retrieval of lab or x-ray results. It is also likely that EHRs will have an electronic problem summary list that outlines a patient's major illnesses, surgeries, allergies and medications. How many times does a physician open a large paper chart, only to have loose lab results fall out? How many times does a physician re-order a test because the results or the chart is missing? It is important to note that paper charts are missing as much as 25% of the time, according to one study.[9] Even if the chart is available; specifics are missing in 13.6% of patient encounters according to another study. Table 2.1 shows the types of missing information and its frequency.[10]

Information type	% Visits information is missing
Lab results	45
Letters/dictations	39
Radiology results	28
History and physical exams	27
Pathology results	15

Table 2.1. Types and frequencies of missing information

According to the President's Information Technology Advisory Committee, 20% of laboratory tests are requested because previous studies are not accessible.[11] This statistic has great patient safety, productivity and financial implications. EHRs allow easy navigation through the entire medical history of a patient. Instead of asking to pull paper chart volume 1 of 3 to search for a lab result, it is simply a matter of a few mouse clicks. Another important advantage is the fact that the record is available 24 hours a day, 7 days a week and doesn't require an employee to pull the chart nor extra space to store it. Adoption of electronic health records has saved money by decreasing full time equivalents (FTEs) and converting records rooms into more productive space, such as exam rooms. Importantly, electronic health records are accessible to multiple healthcare workers at the same time, at multiple locations. Patient information should be available to physicians on call so they can review records on patients who are not in their panel. Also, while a billing clerk is looking at the electronic chart, the primary care physician and a specialist can be analyzing clinical information simultaneously. Furthermore, it is believed that electronic health records improve the level of coding. Do clinicians routinely submit a lower level of care because they know that patient notes are short and incomplete? Templates may help remind clinicians to add more history or details of the physical exam, thus justifying a higher level of coding. (Templates are disease specific electronic forms that essentially allow you to point and click a history and physical exam). Lastly, EHRs provide clinical decision support such as alerts and reminders, which we will cover later in this chapter.

2. The need for improved efficiency and productivity. The goal is to have patient information available to anyone who needs it, when they need it and where they need it. If lab or x-ray results are frequently missing, the implication is that they need to be repeated which adds to this country's staggering healthcare bill. The same could be said for duplicate prescriptions. It is estimated that 31% of the US's $1.9 trillion dollar healthcare bill is for administration.[12] EHRs are more efficient because they reduce redundant paper work and have the capability of interfacing with a billing program that submits claims electronically. Consider what it takes to simply get the results of a lab test back to a patient using the old system. This might involve a front office clerk, a nurse and a physician. The end result is frequently placing the patient on hold or playing "telephone tag". With an EHR, lab results can be forwarded via secure messaging. Electronic health records can help with productivity if templates are used judiciously. As noted, they allow for point and click histories and physical exams, thus saving time. Embedded educational content is one of the newest features of a comprehensive EHR. Clinical practice guidelines, linked educational content and patient handouts can be part of the EHR. This may permit finding the answer to a question while the patient is still in the exam room. Several EHR companies also offer a centralized area for all physician approvals and signatures of lab work, prescriptions, etc. This should improve workflow by avoiding the need to pull multiple charts or enter multiple EHR modules.

3. Quality of care and patient safety. As we have previously suggested, an EHR should improve patient safety through many mechanisms: (a) Improved legibility of clinical notes (b) Improved access anytime and anywhere (c) Reduced duplication (d) Reminders that tests or preventive services are overdue (e) Clinical decision support that reminds us of patient allergies, the correct dosage of drugs, etc. (f) Electronic problem summary lists provide prior diagnoses, allergies and surgeries at a glance. Large healthcare organizations can analyze patient data to improve quality and patient safety. As an example, the healthcare organization Kaiser-Permanente determined that the drug Vioxx had an increased risk of cardiovascular events before that information was published elsewhere.[13] Similarly, within 90 minutes of learning of the withdrawal of Vioxx from the market, the Cleveland Clinic queried its EHR to see which patients were on the drug. Within 7 hours they deactivated prescriptions and notified clinicians via e-mail.[14] This will be discussed in more detail in the patient safety chapter.

4. Public expectations. According to a 2006 Harris Interactive Poll for the Wall Street Journal Online, 55% of adults thought an EHR would decrease medical errors; 60% thought an EHR would reduce healthcare costs and 54% thought that the use of an EHR would influence their decision about selecting a personal physician.[15] The Center for Health Information Technology would argue that EHR adoption results in better customer satisfaction through fewer lost charts, faster refills and improved delivery of patient education material.[16]

5. Governmental expectations. More legislation is being introduced to promote electronic health records. It is the goal of the US Government to have an interoperable electronic health record within the next decade. Grant monies have been made available to encourage adoption of this new technology. Many organizations state that we need to move from the "cow path" to the "information highway". Medicare is acutely aware of the potential benefits of EHRs to help coordinate disease management in older patients.

6. Financial savings. The Center for Information Technology Leadership (CITL) has suggested that computerized order entry in the ambulatory setting would save $44 billion yearly and eliminate more than $10 in rejected claims per patient per outpatient visit.[17] This organization concludes that not only would there be savings from eliminated chart rooms and record clerks; there would be a reduction in the need for transcription. There would also be fewer callbacks from pharmacists with electronic prescribing. It is likely that copying expenses and labor costs would be reduced with EHRs. More rapid retrieval of lab and x-ray reports results in time saving as does the use of templates. More efficient patient encounters means more patients could be seen per day. Better medication management is possible with reminders to use the "drug of choice" and generics.

7. Technological Advances. The timing seems to be right for electronic records partly because the technology has evolved. The Internet and World Wide Web make the application service provider (ASP) concept for an electronic health record possible. An ASP option means that the EHR software and patient data reside on a remote server that you access from the office. Computer speed and memory and bandwidth have advanced such that digital imaging is a reality. Wireless and mobile technologies permit access to the hospital information system, the electronic health record and the Internet using a personal digital assistant or tablet PC.

8. Older and more complicated patients require more coordinated care. According to a 2002 Gallup poll it is extremely common for older patients to have more than one physician as evidenced by the following statistics:
- No physicians—3%
- 1 physician—16%
- 2 physicians—26%
- 3 physicians—23%
- 4 physicians—15%
- 5 physicians—6%
- 6+ physicians—11% [18]

Having more than one physician mandates good communication between the primary care physician, the specialist and the patient. This becomes even more of an issue when different healthcare systems are involved. A 2000 Harris

Interactive survey reported that physicians understand that adverse outcomes result from poor care coordination with chronically ill patients.[19] Nevertheless, a survey of patients with chronic conditions showed that 18% of the population received duplicate tests/procedures, 17% received conflicting information from other clinicians and 14% received different diagnoses from different physicians.[20] In the future, electronic health records will be integrated with regional health information networks so that inpatient and outpatient encounter notes can be accessed and shared, thus improving communication between healthcare entities.

Electronic Health Record Key Components

The following components are desirable in any EHR system. The reality is that many EHRs do not currently have all of these functions.

- Clinical Decision Support Systems (CDSS) to include alerts, reminders and clinical practice guidelines. CDSS is part of computerized physician order entry (CPOE). This will be discussed in more detail in this chapter and the patient safety chapter
- Secure messaging for communication between patients and office staff and among office staff. Telephone triage capability is important
- An interface with practice management software, scheduling software and patient portal (if present). Feature will handle billing and benefits determination
- Managed care module for physician and site profiling. This includes the ability to track Health plan Employer Data and Information Set (HEDIS) or similar measurements and basic cost analyses
- Referral management feature
- Retrieval of lab and x-ray reports electronically
- Retrieval of prior encounters and medication history
- Computerized Physician Order Entry (CPOE). Primarily used for inpatient order entry but ambulatory CPOE also important. This will be discussed in more detail later in this chapter
- Electronic patient encounter. One of the most attractive features is the ability to create and store a patient encounter. In seconds you can view the last encounter and what treatment was rendered
- Multiple ways to input text into the encounter should be available: free text, dictation, voice recognition and templates
- The ability to input or access information via a PDA or tablet PC
- Remote access from the office or home
- E-prescribing
- Integration with a Picture Archiving and Communication System (PACS)
- Knowledge resources for physician and patient, embedded or linked
- Public health reporting and tracking
- Problem summary list that is customizable and includes the major aspects of care: diagnoses, allergies, surgeries and medications

- Ability to scan in text or use optical character recognition (OCR)
- Ability to perform evaluation and management (E & M) coding
- Ability to create graphs or flow sheets of lab results or vital signs
- Ability to create electronic patient lists or disease registries
- Preventive medicine tracking that links to clinical practice guidelines
- Security compliant with HIPAA standards
- Backup systems in place
- Support for client server or web based application service provider option (ASP) [21]

Clinical Decision Support Systems (CDSS)

Traditionally, CDSS meant computerized drug alerts and reminders as part of computerized physician order entry (CPOE) applications (discussed in the next section). Most of the studies in the literature evaluated those two functions. However, according to Hunt, CDSS is "any software designed to directly aid in clinical decision making in which characteristics of individual patients are matched to a computerized knowledge base for the purpose of generating patient specific assessments or recommendations that are then presented to clinicians for consideration".[22] CDSS therefore, should have a broader definition than just alerts and reminders. The table below outlines some of the clinical decision support available today. Calculators, knowledge bases and differential diagnoses programs are primarily standalone programs but they are slowing being integrated into most EHR systems.

Type of Support	Tool Examples
Knowledge	iConsult®, Theradoc®
Calculators	MedCalc 3000 Connect®
Trending/Patient tracking	Flow sheets, graphs
Medications	CPOE and drug alerts
Order sets/protocols	CPGs and order sets
Reminders	Mammogram due
Differential diagnosis	Dxplain®

Table 2.2. Clinical Decision Support

1. Knowledge support. Numerous digital medical resources are being integrated with EHRs. As an example, the American College of Physician's PIER resource is integrated into Allscript's Touch Chart.[23] UpToDate is now available in General Electric's Centricity EHR. [24] IConsult (offered by Elsevier) is a primary care information database available for integration into EHRs. Diagnostic (ICD-9) codes can be hyperlinked to further information or you can use *infobuttons*. Figure 2.3 shows an example of iConsult integrated with the Epic EHR.[25]

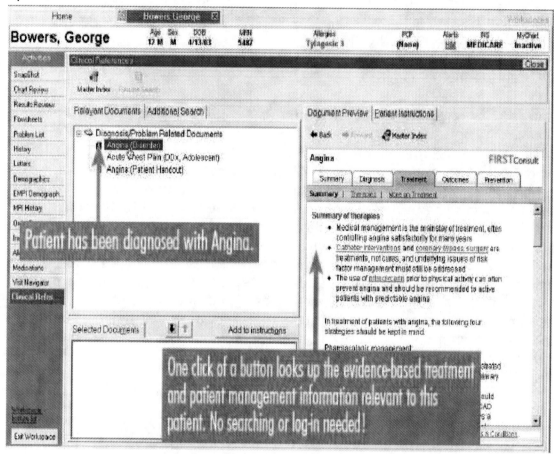

Figure 2.3. iConsult integrated with Epic EHR (courtesy iConsult)

Another interesting integrated knowledge program is the Theradoc Antibiotic Assistant. The program integrates with an EHR's lab, pharmacy and radiology sections to make suggestions as to the antibiotic of choice with multiple alerts.

Clinicians can be alerted via cell phones, pagers or e-mail. Other modules include ADE Assistant, Infection Control Assistant and Clinical Alerts Assistant.[26] A study in the New England Journal of Medicine (NEJM) using this product showed considerable improvement in the prescription of appropriate antibiotics resulting in cost saving, reduced length of stay and fewer adverse drug events.[27]

2. Calculators. It is likely with time that more calculators will be embedded into the EHR, particularly in the medication and lab ordering sections. Figure 2.4 is an example of one of over 100 calculators available on Medcalc3000 as a standalone web and Pocket PC based program. They now offer a "Connect" option that will integrate with EHRs by linking calculators, clinical criteria tools and decision trees.[28]

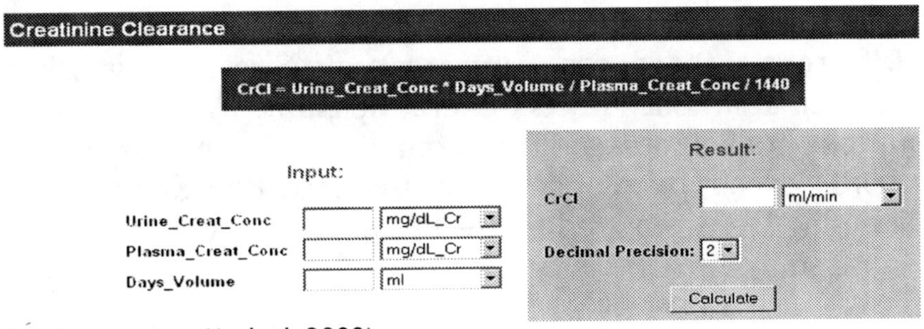

Figure 2.4. Medcalc3000 (courtesy Medcalc3000)

3. Flow sheets, graphs, patient lists and registries. The ability to track and trend lab results and vital signs in, for example, diabetic patients will greatly assist in their care. Furthermore, the ability to use a patient list to contact every patient taking a recalled drug will improve patient safety. Registries will be covered in more detail in the Disease Management chapter.

4. Medication ordering support. Many aspects of medication safety and alerts will be covered in the Patient Safety chapter. Decision support as part of CPOE possesses several rules engines to detect known allergies, drug-drug interactions and excessive dosages. As EHRs and CPOE mature, they will factor in the age, gender, weight, kidney (renal) and liver (hepatic) function of the patient, known contraindications based on known diagnoses as well as the pregnancy and lactation status. Incorporation of these more robust features is complicated and best implemented at medical centers with an established track record of development of EHRs, CDSS and CPOE. As has been pointed out, there are programs that improve antibiotic ordering based on data residing in the EHR.[29]

5. Reminders This will be discussed in greater detail in the Patient safety chapter. Computerized reminders that are part of the EHR assist in tracking the yearly preventative health screening measures, such as mammograms. In a well designed system it should allow for some customization of the reminders as national recommendations change. Reminders are not always heeded by busy clinicians who may choose to ignore them. Preventive measures could be reviewed by the office nurse and overdue tests ordered prior to the visit with the physician.

6. Order sets and protocols. Order sets are groups of pre-established orders that are related to a symptom or diagnoses. For instance, you can create an order set for low back pain that might include ibuprofen and a muscle relaxant that saves keystrokes and time. Order sets can also reflect best practices (clinical practice guidelines), thus offering better and less expensive care. Over one hundred clinical practice guidelines are incorporated into the electronic health record at Vanderbilt Medical Center.[30] An excellent 2007 review of order sets as part of CDSS was reported in the Journal of the American Medical Informatics Association. [31]

7. Differential Diagnoses. Dxplain is a differential diagnosis program developed at Massachusetts General Hospital. When you input the patient's symptoms it generates a differential diagnosis (the diagnostic possibilities). The program has been in development since 1984 and is currently web based. A licensing fee is required to use this program. At this time it can not be integrated into an EHR.[32]

How well clinicians use CDSS programs such as these remains to be seen. They will have to be intelligently designed and rigorously tested in order to be accepted. For more information on CDSS, we refer you to these references. [33-36]

Computerized Physician Order Entry (CPOE)

CPOE is a computer application that processes orders for medications, lab tests, x-rays, consults and other diagnostic tests. Many organizations like the Institute of Medicine and Leapfrog see CPOE as a powerful instrument of change. There is evidence that CPOE will:

1. Reduce medication errors. CPOE has the potential to reduce medication errors through a variety of mechanisms.[37] Since it is electronic you can embed rules-engines that allow for checking allergies, contraindications and other alerts. Koppel et al [38] lists the following advantages of CPOE compared to paper-based systems:
- Overcome the issue of illegibility
- Fewer errors associated with similar drug names
- More easily integrated with decision support systems
- Easily linked to drug-drug interaction warnings
- More likely to identify the prescribing physician
- Able to link to adverse drug event (ADE) reporting systems
- Able to avoid medication errors like trailing zeros
- Available for data analysis
- Can point out treatment and drugs of choice
- Reduce under and over-prescribing
- Prescriptions reach the pharmacy quicker [39]

2. Reduce costs. Several studies have shown reduced length of stay and overall costs in addition to decreased medication costs.[40]

3. Reduce variation of care. One study showed excellent compliance by the medical staff when the drug of choice was changed using decision support reminders.[41] Study conclusions should be interpreted with some note of caution. Many of the studies were conducted at medical centers with well established medical informatics programs where the acceptance level of new technology is unusually high. Several of these institutions like Brigham and Women's Hospital developed their own EHR and CPOE software. Compare this experience with that of a rural hospital that is trying CPOE for the first time with inadequate IT, financial and leadership support.

On the surface CPOE seems easy: just replace paper orders with an electronic format. The reality is that CPOE represents a significant change in workflow and not just new technology. An often repeated phrase is "it's not about the software", meaning, regardless which software program you purchase, it requires a change in the way you do business or workflow.

Adoption of CPOE has been slow, partly because of cost and partly because workflow is slower than scribbling on paper. In a 2006 article based on survey data it was estimated that 24% of US physicians used an ambulatory EHR, as opposed to 5% of hospitals using CPOE.[42] Although physicians have been upset by any new change that does not shorten their work day, many authorities feel EHRs greatly improve many hospital functions. There has been less resistance traditionally in teaching hospitals with a track record of good informatics support. It does require great forethought, leadership, planning, training and the use of physician champions in order for CPOE to work. According to some, CPOE should be the last module of an EHR to be turned on and alerts should also be phased in to bring about change more gradually. Others have recognized nurses as more accepting of change and willing to teach docs "one on one" on the wards. For more information on CPOE we refer you to references.[43-44] Note that CPOE will also be further discussed in the chapter on patient safety.

Electronic Health Record examples

Let's start by mentioning some less expensive choices, available in 2007 that offer a moderate amount of functionality.

Medical Office Online. This web based program is priced at $275/physician per month, plus a user license of $15 per month and an initial set up fee of $675 to import data. The main features offered include:
- Patient demographics
- The ability to scan and attach insurance cards, photo IDs, images and attachments
- Scheduling
- Automatic CPT and ICD-9 charge capture

- Evaluation and Management (E & M) coding
- Prescription generation
- Letter templates
- Refill requests
- Charge capture, accounts receivable, electronic claims submission and billing [45]

SoapWare. Patient encounter data can be inputted by templates or voice recognition. Software resides on either the office PC or a SQL (sequel) server and an ASP was added in the late 2006 timeframe. Standard program features patient handouts, phone messaging, vital sign tracking and the ability to print "superbills". Software has 23 optional modules described in detail on their web site. Three bundled packages are available with some or all of the additional modules for $999-$3999. There is an extra $300 charge for each license and $300 for support/license. This program is very popular among small primary care practices but also covers more than 60 specialties and exists in more than 30+ countries. [46]

VistA Office EHR. This program is basically the same EHR the Veterans Affairs (VA) system uses without the inpatient function. It was released in October 2005 as a joint venture between the VA and the Centers for Medicare and Medicaid Services to promote a more affordable EHR for the US and third world countries. It was initially touted as being free but in reality there is a $2700 licensing fee per physician as well as installation, maintenance and licensing fees for CPT and the software program Cache´. EHR features and a demo are available on the web site. Lab and billing/practice management programs can be added later but will require a MUMPS programmer. VistA-Office is technically not an open source software program but is instead public domain software, which means you can obtain and alter the source code. Medsphere OpenVista is a vendor that will offer an open source version as well as a web based or ASP version. Their program is offered in two configurations: enterprise for an organization and clinic configuration for smaller clinics and multi-specialty groups. For several reasons, this new outpatient initiative should be considered a beta-test program. [47-48]

Other similar less expensive EHR programs:
- Amazing Charts www.AmazingCharts.com
- iGreat www.igreat.com
- MedComSoft www.medcomsoft.com
- Praxis EMR www.praxisemr.com
- E-MDS www.e-mds.com
- SpringCharts www.springcharts.com
- Cottage Med http://mtdata.com/~drred/
- ComChart www.comchart.com

More comprehensive EHR examples

Armed Forces Health Longitudinal Technology Application (AHLTA). This is the program (previously known as CHCS II) that is being deployed worldwide for every Department of Defense (DOD) medical facility (9.2 million patients). It began as CHCS I in about 1992 and was used primarily for outpatient order entry and the ability to retrieve lab, x-ray and pharmacy results. CHCS II is more robust than CHCS I with the ability to use templates to input information into a patient encounter. A unique feature of this EHR is the use of the MEDCIN terminology engine. MEDCIN is a collection of 200,000 terms you can select to construct a history and physical exam. This was selected in order to create a structured note, so each data field can be saved and retrieved and the information can build an Evaluation and Management (E & M) code. The limitation is that it is very time consuming to learn to use this technique and then create your own list of customized profiles. Specialists are used to dictating a very comprehensive note, whereas the use of templates tends to generate very short cryptic notes. Needless to say, templates and MEDCIN have their advocates and their skeptics. Below in Figure 2.5 is an example of a MEDCIN template where you search for the correct terms to input into the patient note. AHLTA has an extensive CPOE application for ambulatory care only. Note the features listed in the left pane below. Other limitations are the lack of interoperability with the VA system and the inability to input/scan documents into the EHR. Excellent demos can be found on a training web site, customized for "providers", nurses, support/techs and clerks.[49-50]

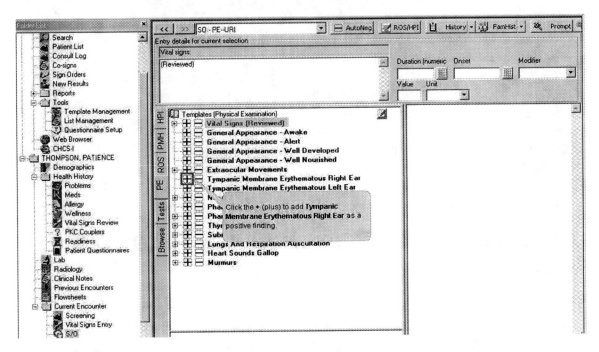

Figure 2.5. AHLTA and MEDCIN terminology engine (adapted from the Naval Medical Center Portsmouth web site)

Veterans Health Administration Computerized Patient Record System (CPRS)
In 1996, the VA introduced *VISTA* (Veterans Health Information Systems and Technology Architecture). CPRS is the Veterans EHR system that serves approximately 7.5 million enrollees and is interoperable with other VA facilities. CPOE accounts for 93% of all prescriptions and the VA system processes approximately 860,000 orders daily. Image archiving and bar coded medication administration is part of CPRS. This EHR is currently used in outpatient, inpatient, Mental Health, intensive care unit (ICU), Emergency Department, Clinic, Homecare, Nursing Home and other settings. Unfortunately, it was built on an old MUMPS programming like CHCS I but will eventually migrate to a more modern architecture and will be known as Health<u>e</u>Vet-Vista. The program has a new patient portal My Health<u>e</u>Vet that allows patients to enter lab and vital signs and create a personal health record. [51-54]

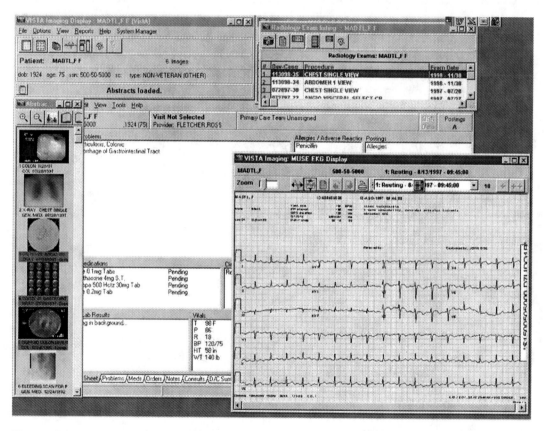

Figure 2.6. VISTA CPRS (courtesy Veterans Affairs)

Specialty EHR. The EHR vendor NexGen offers 15 sub-specialty patient encounter modules. A customized EHR for sub-specialties makes sense because their needs are more narrow and different than those of a primary care physician. An excellent example is their Ophthalmology module that organizes the EHR into a very logical order for eye physicians. Their product is available as a client- server or ASP option. The history and physical exam can be specialty specific and images can be stored in each patient's record. Coding is automatically generated as are letters back to the referring physician. A PDA version is also available. They also offer practice management, document scanning, patient portal and a community health modules (central data repository). [55-56]

Examples of EHR successes and failures

Duke University Medical Center
- EHR developed in 1986
- Total time per patient visit was reduced by 13%
- Pre-exam functions were reduced
- Fewer overlooked problems
- Fewer charting errors
- Fewer prescription errors
- Physician's actual time with patient unchanged [57]

Central Utah Multi-specialty Clinic
- 59 physicians practicing in nine locations
- Clinic used Allscripts Touchworks EHR
- In first year they experienced $900,000 profit due to increased revenue and decreased expenses
- They anticipate savings of $8.3 million over next 5 years [58]

Maimonides Medical Center
- With a new EHR system medication discrepancies have fallen by 60%
- 165,000 potential drug interactions were detected (unknown how many were truly serious) in one year resulting in 82,000 treatment changes
- System used by 100% of medical staff [59]

Cincinnati Children's Hospital Medical Center
- Partial EHR based on Siemens software implemented at a cost of $14 million dollars
- Medication ordering/dispensing errors reduced from 120 to 90 per month
- Program reduced time to get drugs from pharmacy to bedside in half [60]

Cedar-Sinai Hospital
- $34 million CPOE system
- Shut down 3 months after implementation because:
    o System too slow and too many technical problems
    o Poor physician input
    o No phase-in. Mandated CPOE from the beginning [61]

Barriers to EHR Adoption

According to Shortliffe [62], there are four historic constraints to EHR adoption: the need for standardized clinical terminology; privacy, confidentiality and security concerns; challenges to data entry by physicians and difficulties with integrating with other systems. In reality, multiple additional barriers exist to include:

1. Physician resistance. In a monograph by Dr. David Brailer, lack of support by medical staff is consistently the second most commonly perceived obstacle to adoption, behind lack of resources. [63] They have to be shown a new technology makes money, saves time or is good for their patients. None of these can be proven for certain for every practice. Although you should not expect to implement CPOE or go paperless from the beginning, at some point it can no longer be optional. It seems clear that CPOE does take longer than written orders but offers multiple advantages over paper as pointed out previously. One systematic review suggests that documentation time for CPOE was far greater for physicians than nurses. [64] Initial productivity will slow down in the office for 3-12 months which is difficult for clinicians, even though their productivity will likely improve above baseline in the future. Too many studies looking at this topic are based on surveys or theoretical models and not excellent studies. Implementation will not fix old workflow issues and will not work if several physicians in a group are opposed to going electronic.

2. Financial Barriers. Although there are models that suggest significant savings after the implementation of ambulatory EHRs, the reality is that it is expensive. Surveys by the Medical Records Institute, MGMA and HIMSS report that lack of funding is the number one barrier to EHR adoption, cited by about 50% of respondents. [65] In a 2005 study published in *Health Affairs*, initial EHR costs averaged $44 K (range $14-63K) per FTE (full time equivalent) and ongoing costs of $8.5 K per FTE. Financial benefits averaged about $33,000 per FTE provider per year. Importantly, more than half of the benefit derived was from improved coding! This study looked at fourteen primary care practices using two well known EHRs. The average practice showed a return on investment in only 2.5 years. [66] This is not a surprise given the fact that studies have shown that physicians often "under code" for fear of punishment or lack of understanding what it takes to code to a certain level. [67] Keep in mind that integration with other disparate systems such as practice management systems can be very expensive and hard to factor into a cost-benefit analyses. The web based application server provider (ASP) option is less expensive in the short term but unclear in the long term. Many authorities believe that the federal government and insurers will need to add incentives if we are to have universal EHRs, particularly in small physician and/or rural practices. According to a survey by the Commonwealth Fund, adoption of EHR and CPOE was far higher in large physician practices.[68]

3. Loss of productivity. It is likely physicians will have to work at reduced capacity for several weeks to months with gradual improvement depending on training, aptitude, etc. This is a period when physician champions can help maintain momentum with a positive attitude.

4. Work Flow changes. Everyone in the office will have to change the way they route information compared to the old paper system. If planning was well done in advance you should know how your workflow will change. Initially, you will have to maintain a dual system of paper and EHR. Workflow will also determine where you will place computer terminals. Importantly, clinicians will have to maintain eye contact as often as possible and learn to incorporate the EHR in the average patient visit. Use of a movable monitor or tablet PC may help diminish the time the clinician spends not looking at the patient.

5. Integration with other systems. Hopefully, integration with other systems like the practice management software was already solved prior to implementation. Be prepared to pay significantly for programmers to integrate a new EHR with an old legacy system. Most office and hospitals have multiple old legacy systems that do not talk to each other. Systems are often purchased from different vendors and written in different programming languages. It is now more popular to purchase an EHR already integrated with practice management, billing and scheduling software programs.

6. Lack of standards. One can assume that an EHR purchased today will not communicate with other EHRs, although vendors are being pressured to make their products interoperable. The Certification Commission for Healthcare Information Technology (CCHIT) was established and by November 2006 thirty three ambulatory EHRs became certified. The process is expensive and true interoperability is not required at this time.[69] The Department of Health and Human Services department has designated CCHIT as the first recognized certification body. They require certification one year prior to hospitals being allowed to give EHR software to physicians. [70]

7. Adverse legislation. There is concern that previously passed legislation will make it difficult for hospitals and physicians to combine forces and create information networks. The infamous Stark Law prohibits a physician from referring Medicare patients to an entity if he/she has a financial relationship with the entity. The Anti-kickback Act makes it illegal for an individual or entity to offer remuneration of any kind to another individual or entity for referring a patient. It is illegal to have to purchase or lease any covered item or service. In 2006 "safe harbor" exceptions became law and allowed hospitals to provide electronic health records to physicians. The software must be interoperable as defined by a certifying organization (CCHIT), "no more than 12 months prior to the date it is provided to the physician". In addition, the physician pays 15% of the donors cost for the equipment.[71-72] It is unclear if these exceptions will result in increased use of EHRs.

8. Inadequate proof of benefit. Although there is plenty of hype regarding the benefits of EHRs the reality is we need better research. A systematic review by Hunt showed that the effects of clinical decision support systems as an example have not been adequately studied.[22] There have been several studies in 2005 that have shown increased errors as a result of implementing CPOE. [37,73-75] Most of the studies have been criticized for one reason or another but it should come as no surprise that any new technology will create new possibilities for errors. Eventually with better training or re-design they are likely to be overcome.

Practical Tips to Selecting an Electronic Health Record

There is a tendency to pick a well know EHR vendor and hope for the best, much like picking an automobile. Unfortunately the process is far more complex and less dependent on the vendor selected. More important are the specific needs of the group, implementation, training, integration and buy-in by all staff. One early decision that must be made is whether you want to purchase a standard EHR package which means having the software on your own computers. The other choice is an application service provider (ASP) which means a remote server hosts the EHR software and your patient data. Each has its merits and shortcomings.[76]

Features of an ASP Model: [21]
- Charges monthly fees to provide access to patient data on a remote server owned by a vendor. Fees will usually include maintenance, software upgrades, data backups and help desk support
- Lower start up costs
- ASP maintains and updates software
- ASP charges fixed amount regardless of the number of users
- ASP can be completely web based or can require a small software program (thin client) to help share processing tasks
- Requires very little local tech support, thus saves money. Often a better choice for small practices with less IT support
- Must have fast internet connection; should be cable modem, DSL or T1 line
- Enables remote log-ons
- Lease agreement commitments range from 1-5 years
- An excellent free monograph on the ASP model is available online [77]
- Cons:
  - If your ISP is not working, you aren't either
  - Concerns about security and HIPAA

- ○ Concerns about who owns the data. In 2006 a group of physicians refused to pay for a 400% increase in tech support. As a result they were denied access to their patient's records [78]
- ○ Cost of monthly cable fees

An informal survey of EHR companies revealed that 37 out of 72 vendors listed an ASP alternative.[79] You are now also seeing larger practices or hospitals hosting EHR ASP services for smaller practices to make it affordable.[80] An excellent review of ASPs was released in October 2006 by the California HealthCare Foundation. [81]

General questions to vendors

- How many licenses have been sold overall?
- Number of years in business of selling EHRs?
- Number of employees and salesmen?
- Does the company focus primarily on a certain size practice?
- Is there a problem with multiple log-ons at the same time?
- Does the EHR interface with other electronic systems?
- Does the maintenance fee cover travel and does it cover nights, weekends and holidays?
- How much for software upgrades?
- What is the training time required to become truly operational?
- Training cost per user or practice?
- On-site training available?
- Hardware and software requirements? [21]

Twelve Step Method to purchase an EHR

1. List priorities for the practice. Are you trying to save time or money or do you just want electronic patient encounters? Now is the time to study workflow and see how it will change your practice. This is when frequent conferences with your front office staff will be critical to get their input about the processes that need to improve.
2. List the EHR features that you feel will meet your needs.
3. Factor in your future requirements. Do you plan more partners or offices or specialties?
4. Write a simple "request for proposal" (RFP). This will cause you to put on paper all of your requirements. Each vendor will need to respond in writing how they plan to address them. Exact pricing should be part of the RFP. Sample RFPs are available on the Web. [82]
5. Make sure physicians are committed to using the EHR. Look for at least one physician champion and be sure your staff is onboard.
6. Choose the method of inputting: keyboard, mouse, stylus, touch screen or voice recognition?

7. Use practice scenarios to test the vendors to be sure their product will perform as desired.
8. Obtain several references from each vendor and visit each practice if possible. Be sure to select similar practices to yours. The following reference provides an EHR demonstration rating form, questions to ask EHR references and a vendor rating tool. [83]
9. Create a scoring matrix to compare vendors. Look at several EHR rating sources. Keep in mind that much of the survey data is self reported by vendors. Nevertheless, the surveys provide valuable information; often difficult to obtain from a web search. [84-86] There are also several fee-based sites to compare EHR products. The following reference has sections for software, interfaces, third party software, conversion services, implementation services, training services, data recovery services, annual support and maintenance, financing alternatives and terms. It also includes red flags and FAQ's. [87]

10. Decide on software purchase versus ASP option. Survey hardware and network needs.
11. Obtain in writing commitments for implementation and technical support, including data conversion from paper records; interfacing with practice management (PM) software; exact schedule and timeline for training. If you are unhappy with your practice management (PM) software consider buying an EHR-PM combination that you know will work.
12. Take advantage of falling prices and increased competition.

For a recent real world study of EHR implementation by a group of four Internists reported in the literature, see reference. [88] According to one author:

> *"Despite the difficulties and expense of implementing the electronic health record, none of us would go back to paper"*

Inpatient CPOE Implementation Issues. Here are some helpful steps to implement CPOE from hospitals who have made the transition [89]
1. Be sure the entire organization is on board; including management.
2. Pay physician champions. The process may require their services for 1-3 years.
3. Analyze your workflow ahead of time.
4. Create adequate order sets. Several of the hospitals writing this report have over 300 order sets. Order sets should be approved and evidence based.
5. Recognize the politics and leaders to make implementation happen.
6. Set a deadline but be sure it is realistic. You may need dual paper and electronic systems for awhile.

7. Train extensively including all shifts and weekend workers. "Pizza meetings" and nurse champions may be necessary to educate and train recalcitrant physicians. Refresher training needs to take place in the workplace.
8. Physician resistance may require one-on-one meetings with a physician champion.
9. Publish any quality improvement that occurs as a result of EHR implementation. The benefits of CPOE seem to be more obvious to pharmacists and nurses, than to physicians.
10. "Crack the Whip". This would only apply to the rare physician who refuses to use CPOE, in the face of widespread acceptance.

For further reading on electronic health records we refer you to the UWF resource site at www.uwf.edu/sahls/medicalinformatics/

Conclusion

In spite of the slow acceptance of EHRs by clinicians, they continue to proliferate and improve over time. As more studies show reasonable return on investment, medical groups that have been sitting on the fence will make the financial commitment. Also, as more practices purchase an EHR, the average price for a system will decrease. The future will include better integration and interoperability and features to support initiatives such as "pay for performance". It is unclear when we will see a universal EHR at all acute and chronic inpatient and outpatient facilities. As a new trend, we are seeing outpatient clinicians who have opted to re-engineer their business model based on an EHR. Their goal is to reduce overhead by having fewer support staff and to concentrate on seeing fewer patients but with more time spent per patient.[90]

References

1. Schwartz WB Medicine and the computer. The promise and problems of change NEJM 1970;283:1257-64
2. The Computer-Based Patient Record: An Essential Technology for Health Care, Revised Edition (1997) Institute of Medicine. The National Academies Press www.nap.edu/books/0309055326 (Accessed October 15 2005)
3. Berner ES, Detmer DE, Simborg D Will the wave finally break? A brief view of the adoption of electronic medical records in the United States JAIMA 2004;12
4. Katrina Health Records www.katrinahealth.org (Accessed October 8 2005)
5. Audet AM et al Information Technologies: When will they make it into Physicians Black Bags? 2003 Commonwealth Fund National Survey of Physicians and Quality of Care. Medscape General Medicine 2004;6 www.medscape.com/viewarticle/493210. (Accessed December 7 2004)
6. England's NHS experiencing delays with National IT Program www.ihealthbeat.org April 28 2004 (Accessed January 29 2006)
7. Taylor H, Leitman R European physicians especially in Sweden, Netherlands and Denmark, lead US in use of electronic medical records. Health Care News.

Harris/interactive. Vol 2 (16) 2002. www.harrisinteractive.com (Accessed July 20 2005)

8. Stead WW, Kelly BJ, Kolodner RM Achievable steps toward building a National Health Information Infrastructure in the United States JAMIA 2005;12:113-120

9. Tang PC et al. Measuring the Effects of Reminders for Outpatient influenza Immunizations at the point of clinical opportunity. JAMIA 1999;6:115-121

10. Smith PC et al Missing Clinical Information During Primary Care Visits JAMA 2005;293:565-571

11. The President's Information Technology Advisory Committee (PITAC) http://www.nitrd.gov/pubs/pitac/ (Accessed January 28 2006)

12. Lohr S Building a Medical Data Network. The New York Times. November 22 2004. (Accessed December 20 2005)

13. US Food and Drug Administration http://www.fda.gov/ola/2004/vioxx1118.html (Accessed Aug 15 2006)

14. Badgett R, Mulrow C Using Information Technology to transfer knowledge: A medical institution steps up to the plate [editorial] Ann of Int Med 2005;142;220-221

15. Wall Street Journal Online/Harris Interactive Health-Care Poll www.wsj.com/health (Accessed October 24 2006)

16. Potential benefits of an EHR. AAFP's Center for Health Information Technology. www.centerforhit.org/x1117.xml (Accessed January 23 2006)

17. Center for Information Technology Leadership. CPOE in Ambulatory Care. www.citl.org/research/ACPOE.htm (Accessed November 9 2005)

18. Jacobe D Worried about...the Financial Impact of Serious Illness. Gallup Serious Chronic Illness Survey 2002 http://poll.gallup.com/content/default.aspx?ci=6325&pg=1 (Accessed January 29 2006)

19. *Chronic Illness and Caregiving*, a survey conducted by Harris Interactive, Inc., 2000. www.harrisinteractive.com (Accessed January 29 2006)

20. National Public Engagement Campaign on Chronic Illness-Physician Survey, conducted by Mathematica Policy Research, Inc., 2001. (Accessed October 15 2005)

21. Carter J Selecting an Electronic Medical Records System Practice Management Center. American College of Physicians October 2004. www.acponline.org/pmc (Accessed January 10 2005)

22. Hunt DL et al Effects of Computer-Based Clinical Decision Support Systems on Physician Performance and Patient Outcomes: A systematic review JAMA 1998;280(15);1339-1346

23. American College of Physicians Physician Information and Educational Resource (ACP Pier) http://pier.acponline.org/index.html?hp (Accessed January 28 2006)

24. Hoffheinz F. UpToDate. Personal Communication November 20 2006

25. FirstConsult integration into EHR using iConsult http://iconsult.elsevier.com/demo.html (Accessed December 15 2005)

26. Theradoc Antibiotic Assistant www.theradoc.com (Accessed February 24 2007)

27. Evans RS et al A Computer-Assisted management Program for Antibiotics and Other Anti-infective agents NEJM 1998;338 (4):232-238

28. Introducing MEDCALC 3000 www.mdng.com November 2004 (Accessed December 2004)

29. Bates DW et al A Proposal for Electronic Medical Records in US Primary Care JAMIA 2003;10:1-10

30. Giuse NB et al Evolution of a Mature Clinical Informationist Model JAMIA 2005;12:249-255

31. Bobb AM, Payne TH, Gross PA. Viewpoint: Controversies surrounding use of order sets for clinical decision support in computerized order entry. JAMIA 2007;14:41-47

32. DxPlain. http://www.lcs.mgh.harvard.edu/projects/dxplain.html . (Accessed January 23 2006)

33. Osheroff JA et al Improving Outcomes with Clinicial Decision Support: An Implementer's Guide. HIMSS Publication 2005. www.HIMSS.org
34. Briggs B Decision Support Matures. Health Data Management August 15 2005. www.healthdatamanagement.com/html/current/CurrentIssueStory.cfm?Post ID=19990 (Accessed August 20 2005)
35. M.J. Ball Clinical Decision Support Systems: Theory and Practice. Springer. 1998
36. Clinical Decision Support Systems in Informatics Review. www.informatics-review.com/decision-support. (Accessed January 23 2006)
37. Bates DM, Teich JM, Lee J et al The impact of Computerized Physician Order Entry on Medication Error Prevention JAMIA 1999;6:313-321
38. Koppel R et al Role of Computerized Physician Order Entry Systems in Facilitating Medication Errors. JAMA 2005;293:1197-1203
39. Mekhjian HS Immediate Benefits Realized Following Implementation of Physician Order Entry at an Academic Medical Center JAMIA 2002;9:529-539
40. Tierney WM, Miller ME Physician Inpatient Order Writing on Microcomputer Workstations: Effects on Resource Utilization. JAIMA1993;269:379-383
41. Teich JM et al Toward Cost-Effective, Quality Care: The Brigham Integrated Computing System. Pp19-55. Elaine Steen [ed] The Second Annual Nicholas E. Davis Award: Proceedings of the CPR Recognition symposium. McGraw-Hill 1996
42. Ashish KJ et al. How common are electronic health records in the US? A summary of the evidence. Health Affairs 2006;25:496-507
43. A Primer on Physician Order Entry. California HealthCare Foundation. First Consulting Group September 2000. www.chcf.org (Accessed September 20 2006)
44. Kuperman GJ, Gibson RF Computer Physician Order Entry: Benefits, Costs and Issues Ann Intern Med 2003;139:31-39
45. Medical Office Online. www.medicalofficeonline.com. (Accessed January 24 2006)
46. Soapware EMR www.docs.com (Accessed January 20 2006)
47. VistA-Office HER www.vista-office.org (Accessed September 26 2005)
48. Medshere. www.medsphere.com (Accessed January 5 2007)
49. CHCSII/ALHTA site. http://www-nmcp.med.navy.mil/AHLTA/AHLTA%20Training%20Tools/index.html (Accessed January 25 2006)
50. AHLTA-Electronic Health Records www.ha.osd.mil/AHLTA/ (Accessed December 2 2005)
51. Murff HJ, Kannry J Physician satisfaction with two order entry systems. JAMIA 2001;8:499-509
52. The VA electronic health record http://www.hhs.gov/healthit/attachment_2/iii.html (Accessed January 28 2006)
53. Veterans Health Information System and Technology Architecture (VISTA) http://www1.va.gov/vha_oi/docs/What_is_VistA.pdf (Accessed January 28 2006)
54. From VistA to HealtheVet-Vista. GAO Training week. Kolodner R. http://www1.va.gov/vha_oi/docs/GAO_Education_Week_November_2004.ppt (Accessed January 28 2006)
55. NexGen web site www.nexgen.com (Accessed January 28 2007)
56. Vision Associates. www.visionassociates.net (Accessed January 28 2006)
57. Garrett E W. How the Past teaches the Future JAMIA 2001 May-Jun; 8(3): 222-234
58. Barlow S, Johnson J, Steck J The Economic Effect of Implementing an EMR in an outpatient clinical setting. J. of Healthcare Information Management 2004;18:1-6
59. Maimonides Medical Center www.ihealthbeat.org May 14 2002 (Accessed September 20 2005)
60. Cincinnati Children's Medical Center www.ihealthbeat.org September 11 2003 (Accessed December 12 2004)
61. Chin T. Doctors pull plug on paperless system. www.ama-assn.org/amednews/2003/02/17/bil20217.htm (Accessed March 15 2005)
62. Shortcliffe E The Evolution of electronic medical records Acad Med 1999;74:414-419

63. Brailer DJ, Terasawa EL Use and Adoption of Computer Based Patient Records California Healthcare Foundation 2003 www.chcf.org (Accessed February 10 2005)
64. Poissant L, Pereira J, Tamblyn R et al. The impact of electronic health records on time efficiency of physicians and nurses: a systematic review. JAMIA 2005;12:505-516
65. Wang SJ et al A Cost-Benefit Analysis of Electronic Medical Records in Primary Care Amer J of Med 2003;114:397-403
66. Miller RH et al The Value of Electronic Health Records In solo or small group practices Health Affairs 2005;24:1127-1137
67. King MS, Sharp L, Lipsky M. Accuracy of CPT evaluation and management coding by Family Physicians. J Am Board Fam Pract 2001;14(3):184-192
68. 2003 Commonwealth Fund National Survey of Physicians and Quality of Care. http://www.cmwf.org/surveys/surveys_show.htm?doc_id=278869 (Accessed December 7 2004)
69. Certification Commission for Healthcare Information Technology (CCHIT) www.cchit.org (Accessed November 2 2006)
70. HHS Recognizes Certification Body to Evaluate Electronic Health Records October 26 2006 www.govtech.net (Accessed November 2 2006)
71. Physician Self-Referral Exceptions For Electronic Prescribing and Electronic Health Records Technology. Centers for Medicare and Medicaid Services. www.cms.hhs.gov/apps/media/press/release.asp?Counter=1920 (Accessed November 24 2006)
72. Congressional Budget Office Cost Estimate www.cbo.gov (Accessed November 24 2006)
73. Han YY et al Unexpected Increased Mortality After Implementation of a commercially sold computerized physician order entry system Pediatrics 2005; 116:1506-1512
74. Nebecker JR et al High Rates of Adverse Drug Events in a Highly Computerized Hospital. Arch Int Med 2005; 165:1111-1116
75. Bates DW Computerized physician order entry and medication errors: Finding a balance. J of Bioinform 2005;38:259-261
76. Practice Partner ASP http://www.pmsi.com/asp/asp.htm (Accessed January 29 2006)
77. Physician Practices: Are Application Service Providers Right for You? Prepared for California HealthCare Foundation www.chcf.org October 2006 (Accessed January 3 2007)
78. Multiple doctors cut off from records by Dr. Notes. The Business Journal. www.bizjournals.com (Accessed December 18 2005)
79. EHR Attributes and Subattributes. May 2004 http://www.azdoqit.com/tools/IOM_Gold_Standard_EHR.pdf (Accessed January 29 2006)
80. Havenstein H Emerging ASP model targets health records May 9 2005. Computerworld http://www.computerworld.com/managementtopics/outsourcing/asp/story/0,108 01,101597,00.html (Accessed January 29 2006)
81. Physician Practices: Are Application Service Providers Right for You? www.cfcf.org (Accessed October 20 2006)
82. Electronic Medical Records: A Buyer's Guide for Small Physician Practices. Forrester Research. October 2003. www.chcf.org (Accessed January 29 2006)
83. Adler KG How to Select an Electronic Health Record System. Family Practice Management February 2005. www.aafp.org/fpm (Accessed January 20 2006)
84. EMR Evaluation Tool and User Guide. California HealthCare Foundation. www.chcf.org/topics/view.cfm?itemID=21520 (Accessed January 20 2006)
85. American College of Rheumatology. Electronic Medical Records for the Physician's Office. www.rheumatology.org/products/coding/03emr_ack.asp

(Accessed January 30 2006)
86. AC Group Releases Mid-Year 2005 EHR and EMR survey. Revised January 10 2006. www.acgroup.org (Accessed January 15 2006)
87. eHealth Initiative EHR Master Quotation Guide. eHealth Initiative. http://www.providersedge.com/ehdocs/ehr_articles/eHealth_Initiative-EHR_Master_Quotation_Guide.pdf#search=%22%22eHealth%20Initiative%20EHR%20Master%20Quotation%20Guide%22%20%20%22 (Accessed January 30 2006)
88. Baron RJ, Fabens EL, Schiffman M . Electronic Health Records: Just around the corner? Or over the cliff? Annals of Int Med 2005;143:222-226
89. Baldwin G Bring Order to CPOE with 10 make or break steps (and 5 myths) HealthLeaders Magazine. November 14 2005. www.Healthleaders.com (Accessed December 2 2005)
90. Diamond J, Fera B. Implementing an EHR. 2007 HIMSS Conference February 25-March 1. New Orleans

Chapter 3: Interoperability

Learning Objectives

After reading this chapter the reader should be able to:
- Identify the need for and benefits of interoperability
- Describe the concept of regional health information organizations and how they fit into a national health information network
- State the most important data standards and their role in interoperability
- Describe the importance of data security and privacy as part of HIPAA

The Need for Interoperability

As the name implies, interoperability means that one electronic application or system is able to connect or exchange information with another. This is a critical element in the future success of health information exchange (HIE) at the local, regional and national level. This chapter will discuss three aspects of interoperability: regional health information organizations (RHIOs), data standards and HIPAA regulations.

Interoperability is of great interest to the federal government because of their strategy to create a National Health Information Network (NHIN). To further promote this initiative they created the Commission on Systemic Interoperability (CSI). The Commission is composed of fourteen members tasked with outlining how we will achieve a universal interoperable electronic health record. Its much awaited report was published in October 2005. A summary of the report is as follows:

- **Adoption.** The federal government should develop incentives for physicians and insurers that include grants and pay for performance initiatives. They should also work to revise or eliminate legal barriers such as the Stark and Anti-Kickback laws. Gaps in health IT adoption should be identified and remedial policies should be developed. The shortage in health IT manpower needs to be addressed and corrected. Lastly, the public needs to know that interoperability will ultimately improve the quality of medical care and patient safety. At this point, it is fairly clear that the public and the average physician have little knowledge of the NHIN.

- **Interoperability.** There is a need to certify health IT products in terms of functionality, security and interoperability. Data standards must be developed with the help of the American Health Information Community (AHIC) and the National Committee on Vital and Health Statistics (NCVHS). The AHIC should require HIPAA to cover privacy specific for health IT. Also needed are standards for labels and packaging. The AHIC should establish a phased-in approach for an interoperable drug record for every citizen by 2010.

- **Connectivity.** National standards for patient authentication and identity need to be developed. The Department of Health and Human Services should work with other agencies to fund a national health information network. There should be criminal punishment for privacy violations. Patients should not be discriminated against based on health data [1-2]

The Certification Commission for Healthcare Information Technology (CCHIT) was created by HIMSS, AHIMA and Alliance. Its goals are as follows:
- Reduce the risk of health information technology (HIT) investment by physicians
- Ensure interoperability of HIT
- Enhance the availability of HIT incentives
- Accelerate the adoption of interoperable HIT

Their immediate goal was to certify inpatient and ambulatory electronic health records by June 2006 and that goal has been accomplished.[3] The process costs about $28,000 per vendor and that has brought the ire of many small vendors.[4]

The Health Information Technology Standards Panel (HITSP) was established by the Department of Health and Human Services to build consensus regarding how to achieve the standards necessary for interoperability. They have approved three sets of specifications and forwarded them to the American Health Information Community (AHIC) for review and approval. [5-6]

Regional Health Information Organizations (RHIOs)

Background

In April 2004 President Bush signed Executive order 13335 creating the Office of the National Coordinator for Health Information Technology (ONCHIT) and at the same time called for interoperable electronic health records within the next decade.[7]

In November 2004 (ONCHIT) sent out a "Request for Information" as to how the National Health Information Network (NHIN) should be established. In particular, they wanted to know how the NHIN would be governed, financed, operated and maintained. Five hundred and twelve responses were received and published in a summary paper. The June 2005 report concluded that the

National Health Information Network should:
- "Be a decentralized architecture built using the Internet linked by uniform communications and a software framework of open standards and policies
- Reflect the interests of all stakeholders and be a joint public/private effort
- Be composed of public and private stakeholders who oversee the determination of standards and policies
- Be patient centric with sufficient safeguards to protect the privacy of personal health information
- Have incentives to accelerate deployment and adoption of a NHIN
- Enable existing technologies, federal leadership, prototypes and certification of EHRs
- Address better refined standards, privacy concerns, financing and discordant inter and intrastate laws regarding health information exchange" [8]

The decision was made to form regional non-governmental networks that would eventually be connected to create the NHIN (Figure 3.1). It is unclear if this approach was taken because of the realization of the mammoth effort and expenditure required for a centralized governmental operated NHIN or to allow for local trial and error and maximal input from the various regions and healthcare organizations around the country.

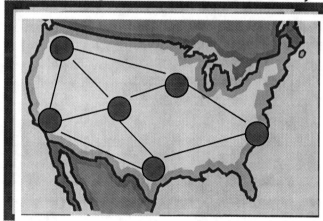

Figure 3.1. NHIN Model   RHIO =

Dr Brailer (ONCHIT) made suggestions as to how RHIOs might proceed:
- Leverage the Internet as an infrastructure; think web-based
- Build upon existing successes; take advantage of any existing infrastructure

- Realistic implementation plan; build "incrementally" or by phases or modules
- Strong MD involvement; involve medical schools and medical societies
- Hospital leadership commitment; much of the information to be shared comes from hospital IT systems
- Do not exclude any stakeholders; RHIOs should consist of multiple types of healthcare organizations
- Inclusion of local public health officials; the goal is to also develop a public health information network or PHIN
- Support from the business community; vendors who have networking experience will be valuable partners
- Neutral managing partner; establish a commission or network authority [9]

The early goals would have to be achievable such as test results retrieval, electronic claims submission, e-prescribing, simple order entry and much later electronic health records. The planning phase for this would likely take several years and would necessitate grant support. Two models of health information exchange (HIE) have appeared and both have their followers:

- Federated—means that data will be stored locally on a server at each entity such as hospital, pharmacy or lab. Data therefore has to be shared among the users of the RHIO
- Centralized—means that the RHIO operates a central data repository that all entities must access

Table 3.1 outlines some of the pros and cons of each model.

	Centralized	Federated
Pros	• Simplicity • Data appearance is uniform • Faster access to data • Easier to create because it is web-based	• Greater privacy • Good examples exist • Buy-in may be easier if data is local
Cons	• Higher hardware costs • Higher operating costs • More difficult with very large RHIOs	• Data display might not be uniform • Data retrieval delays from others

Table 3.1. Pros and Cons of RHIO models (Adapted from Scalese [10])

In order for a RHIO to succeed multiple participants will need to be involved in the planning phase. Examples would be:

- Insurers
- Physicians
- Hospitals
- Medical societies

- Medical schools
- Medical Informatics programs
- State and local government
- Employers
- Consumers
- Pharmacies and pharmacy networks
- Business leaders and selected vendors
- Public Health departments

Multiple functions need to be addressed according to the consulting group HealthAlliant such as:

- Financing: it will be necessary to develop short term start up money and more importantly a long term business plan to maintain the program
- Regulations: what data, privacy and security standards are going to be used?
- Information technology: who will create and maintain the actual network? Who will do the training? Will the RHIO use a centralized or de-centralized data repository?
- Clinical process improvements: what processes will be selected to improve? Claims submission? Who will monitor and report the progress?
- Incentives: other than marketing what incentives exist to have the disparate forces join?
- Public relations (PR): you need a PR division to get the word out regarding the potential benefits of creating a RHIO
- Consumer participation: in addition to the obvious stakeholders you need input from consumers/patients [11]

The expectation is that RHIOs will save money once they are established. It is presumed that the network will reduce office labor and duplication of orders. Table 3.2 shows a model of predicted savings from the Center for Information Technology Leadership (CITL).[12]

	Payer	Clinician	Lab	Radiology	Pharmacy	Public Health	Total
National Savings	$21.6	$24	$13.1	$8.2	$1.3	$0.1	$68.3

Table 3.2. Projected savings (in billions) from RHIO creation (adapted from CITL)

Many people feel that insurers are likely to benefit more from community networks than clinicians. It is important to point out that these are projected models and the predictions may be overly optimistic and not applicable to every RHIO. It is clear that one of the benefits of a health information exchange is more cost effective electronic claims submission. As evidenced by the Utah Health Information Network, a paper claim costs $8, an electronic claim $1 and the charge by the RHIO of 20 cents; therefore a savings of $6.80.[13]

Currently in the United States there are about twenty RHIOs actually exchanging clinical information and about four hundred in the formative stages. It is unknown how many started and failed. The Santa Barbara County Care Data Exchange is a high visibility RHIO that folded in March 2007 due to legal, technological and financial issues. [14] We can only hope that there will be valuable "lessons learned" from this project.

Examples of Regional Health Information Organizations

Utah Health Information Network. Created in 1993, it has been one of the most financially successful non-profit statewide RHIOs in existence. Figure 3.2 demonstrates the diverse entities associated with this RHIO. Their web site is highly educational and includes a fee schedule. [15]

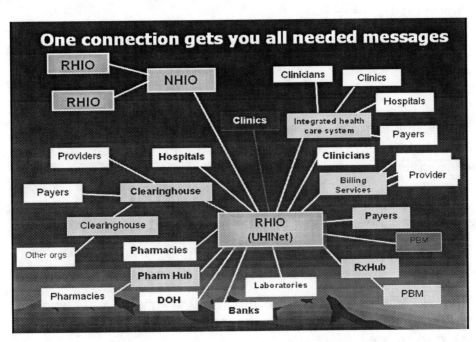

Figure 3.2. Utah Health Information Network (published with permission)

Kentucky Health Electronic Network
- Kentucky is the first state to pass a bill calling for the establishment of a <u>statewide</u> RHIO. It will be interesting to see if using a state model instead of regional model is more successful. Maine is another state that plans a state wide information network by 2010
- The Universities of Kentucky and Louisville will create the Kentucky Health Care Infrastructure Authority
- It will be funded by the state, with efforts to obtain federal and private industry grants from, example, IBM [16]

Indiana Health Information Exchange. Multiple partners helped create this RHIO in 1999 including the Regenstrief Institute that is part of the Indiana University School of Medicine. The RHIO includes twenty one hospitals from five hospital systems and sixty six percent of outpatient physician offices in the Indianapolis area. It is their goal to eventually cover the remainder of Indiana. They also wanted to be an example for the rest of the country, employ more workers and create more data for better research. The network includes the state and local public health departments and the homeless shelters. They realized early on that they would be able to create the following:

- Clinical secure messaging
- Clinical abstracts
- Treatment reminders to physicians
- Physician profiling data
- Results review
- Clinical quality reports
- Research
- Electronic laboratory reports for public health
- Syndromic surveillance (looking for syndromes like flu like illnesses to track epidemics or bioterrorism)
- Adverse Drug Event (ADE) detection
- Improvement in integration efficiencies or inefficiencies
- They plan to launch medication reconciliation, diabetes and cholesterol management and breast cancer and colorectal cancer screening

Although they are considered one of most robust of RHIOs with a successful business plan, they have had several near melt downs due to financial issues.[17]

RHIO Consultants

HealthAlliant. They are a not-for-profit consulting company involved with all phases of RHIOs; planning, formation and financing. So far they have collaborated to create 5 RHIOs:

1. Central Appalachian Health Improvement Partnership (CAHIP) that addresses more remote areas of Tennessee and Virginia
2. Taconic Health Information Network and Community (THINC) in New York State
3. Rhode Island Health Improvement Initiative (RIHII) that also includes an e-prescribing initiative
4. A rural Nebraska healthcare access initiative
5. Choice Regional Health Network located in Olympia, Washington [18]

Healthvision. This organization hosts web-based solutions for all aspects of building RHIOs. Healthvision refers to their RHIOs as Connected Healthcare Communities. Currently they host more than 7.7 million patient records for networks in Texas, Connecticut, Maine, California, Virginia and New York. A centralized model of data storage is used. The organization also has experience with the Continuity of Care Record (CCR) that is a brief summary of care a patient has received as an inpatient or outpatient. This electronic document would be an excellent choice for emergency room visits or hospital transfers or discharges. It might be viewed as an interim solution until all EHRs are interoperable. At that point, the CCR might still be a reasonable "snap shot" of a patient when a physician or healthcare entity just wants a summary of the most significant patient issues. Some have suggested that personal health records be modeled after the CCR. As of 2007 CCR elements and standards are not complete.[19]

Regional Health Information Organization concerns
- Everyone has a different business model. Is this a public utility with no public funding?
- Who will fund networks long term? Insurers? Employers? Consumers? Neither private nor government organizations take full responsibility. The AHRQ funded grants of $18.6 million in late 2005 and will go to only four RHIOs [20]
- What happens when the grant money expires?
- Will we have universal standards or different standards for different RHIOs?
- There could be dependence on vendors. BCBS of Tennessee will partner with Cerner who has an EHR product. Cerner will receive a per member per month subscription fee [21]
- What to do with geographical gaps in RHIOs and what regions should they cover? Should they be based on geography, insurance coverage or prior history?
- Are poorer states and regions at a disadvantage?
- What are the incentives for competing hospitals and their CIOs/CEOs in the average city to truly collaborate?
- Do you need to create a neutral third party to organize the RHIO?

Data Standards

According to the 2003 Institute of Medicine Patient Safety: Achieving a New Standard for Care:
> "One of the key components of a national health information infrastructure will be data standards to make that information understandable to all users" [22]

In order for RHIOs to succeed and EHRs to connect there needs to be a standard language, otherwise you have a "Tower of Babel". We use standards every day but often take them for granted. All languages are based on the rules or standards of grammar. The plumbing and electrical industries depend on standards that work in every state. The railroad industry had to decide many years ago what gauge railroad they would use to connect railroads throughout the United States. Dr. Brailer believed that standards should come first to promote interoperability of electronic health records and not follow adoption of EHRs. Although we have come a long way towards universal standards, we are not there. The progress has been slowed in part due to the fact that these standards determining organizations (SDOs) are voluntary.

The next sections will discuss the major data standards and how the standards facilitate the transmission of data. Not all data standards have been included in the following sections and many standards are still a "work in progress". [23]

Health Level Seven (HL7)
- A not-for-profit standards development organization (SDO) with chapters in 30 countries
- Health Level Seven's domain is clinical and administrative data transmission and perhaps is the most important standard of all
- "Level Seven" refers to the highest level of the International Organization for Standardization (ISO)
- HL7 is a data standard for communication or messages between:
  - Patient administrative systems (PAS)
  - Electronic practice management
  - Lab information systems (interfaces)
  - Dietary
  - Pharmacy (clinical decision support)
  - Billing
  - Electronic health records
- HL7 uses XML (extensible markup language) developed by the World Wide Web Consortium
- HL7 not a software program
- The most current version of the HL7 standard is 3.0 but version 2.0 still widely in use
- HL7 group involved with developing interoperable health records for Katrina victims
- HL7 example:

```
MSH|^~\&|EPIC|EPICADT|SMS|SMSADT|199912271408|CHARRIS|ADT^A04|1817457|
EVN|A04|199912271408|||CHARRIS
PID||0493575^^^2^ID 1|454721||DOE^JOHN^^^^|DOE^JOHN^^^^|19480203|M|
NK1||CONROY^MARI^^^^|SPO||(216)731-4359||EC|||||||||||||||||||||||||
PV1||O|168 ~219~C~PMA^^^^^^^^^||||277^ALLEN FADZL^BONNIE^^^^||||||||||
```

Figure 3.3. HL7 example (the vertical bars are called pipes and separate the bits of data) [24]

Digital Imaging and Communications in Medicine (DICOM)
- Formed by the National Electrical Manufacturers Association (NEMA) and the American College of Radiology who first met in 1983! Perhaps they recognized the future potential of digital x-rays
- As more radiological tests became available digitally, by different vendors, there was a need for a common data standard
- DICOM supports a networked environment using TCP/IP protocol (basic Internet protocol)
- Also applicable to an offline environment [25]

Institute of Electrical and Electronics Engineers (IEEE). IEEE is the organization responsible for writing standards for medical devices. This includes infusion pumps, heart monitors and similar devices. [26]

Logical Observations: Identifiers, names, codes (LOINC)
- Standardized codes for electronic exchange of lab results back to hospitals, clinics and payers
- LOINC database has more than 30,000 codes used for transmitting lab results
- Maintained by the Regenstrief Institute at the Indiana School of Medicine [27]

EHR-Lab Interoperability and Connectivity Standards (ELINCS)
- Created in 2005 as a lab interface for ambulatory EHRs
- Includes
  - Standardized format and content for messages
  - Standardized model for exchanging messages
  - Standardized coding (LOINC)
- Certification Commission for Healthcare Information Technology (CCHIT) has proposed that ELINCS be part of EHR certification
- HL7 plans to adopt and maintain the ELINCS standard
- California Healthcare Foundation is seeking pilot projects to test this standard [28]

RxNorm
- New standard for drugs; developed by the National Library of Medicine
- Includes three drug elements: the active ingredient, the strength and the dose
- The standard for e-prescribing
- Only covers US drugs at this point [29]

National Council for Prescription Drug Programs (NCPDP)
- A standard for exchange of prescription related information
- Facilitates pharmacy related processes
- Standard for <u>billing </u>retail drug sales [30]

Systematized Nomenclature of Medicine: Clinical Terminology (SNOMED-CT)
- Developed by the American College of Pathologists
- SNOMED will be used by the FDA and the Department of Health and Human Services
- Clinical terminology commonly used in software applications including EHRs
- Currently includes about 1,000,000 clinical descriptions
- Terms are divided into 11 axes or categories
- Provides more detail by being able to state condition A is due to condition B
- Links to LOINC and ICD-9. In 2003, the National Library of Medicine paid for free downloads of SNOMED through 2008
- Currently used in over 40 countries
- EHR vendors like Cerner and Epic are incorporating this standard into their EHRs
- Some confusion remains between SNOMED and ICD-9, with the latter being used primarily for billing and the former for communication of clinical conditions [31-32]
- A study at the Mayo Clinic showed that SNOMED-CT was able to accurately describe 92% of the most common patient problems [33]

```
Tuberculosis
D E - 1 4 8 0 0

 .   .   .   .

 .   .   .   .

 .   .   . Tuberculosis
 .   . Bacterial infections
 . E = Infectious or parasitic diseases
D = disease or diagnosis
```

Figure 3.4. SNOMED-CT Example: Tuberculosis

International Classification of Diseases 9th revision (ICD-9)
- Is published by the World Health Organization to allow mortality and morbidity data from different countries to be compared
- Although it is the standard used in billing, it is not ideal for distinct clinical diseases
- ICD-10 was endorsed in 1994 but not used in the US. It will be mandated by the federal government eventually. ICD-10 will provide a more detailed description with 7 rather than 5 digit codes. ICD-10 would result

in about 200,000 codes instead of the 24,000 codes currently associated with ICD-9. A study by Blue Cross Blue Shield estimates that adoption of ICD-10 would cost the US healthcare industry about $14 billion over the next 2-3 years. The more digits included generally results in higher reimbursement.[34-36]

Current Procedural Terminology (CPT)
- Used for billing the level and complexity of service rendered
- Developed, owned and operated by the American Medical Association (AMA) for a fee
- 2003 version contains 8,107 billing codes
- In the case of EHRs, most will convert the diagnosis to an ICD-9 code and help verify the CPT code by automatically adding the elements inputted into the history and physical exam [37]

Health Insurance Portability and Accountability Act (HIPAA)

HIPAA became a reality in 1996 but only in the past several years has it been implemented and taken seriously. Although this subject matter could have been discussed elsewhere, it seemed reasonable to include it under the module on interoperability as it affects healthcare data transmission and storage. In essence, HIPAA has added privacy and security standards for patient data handling similar to the data standards just discussed.

PHYSICIAN FUNNIES

Courtesy California
Medical Association

The public is interested in privacy and security as evidenced by the National Consumer Health Privacy Study of 2005:
- 67% of Americans surveyed are concerned about personal health information privacy
- 52% were concerned about medical information affecting job opportunities
- 98% are willing to share health information with their physician, only 27% with drug companies and 20% with government agencies
- 66% thought paper records more secure than electronic! [38]

Title I of HIPAA deals with
a) the protection of healthcare insurance benefits for workers and families when they change jobs or are discharged from employment
b) a medical savings account project
c) expanded enforcement of fraud and abuse.

We will be discussing only Title II or subtitle F or *Administration Simplification* (when has the government ever simplified anything administrative?). The Administrative Simplification section has four areas of concern:

1. Standardization of electronic patient health, administrative and financial data
2. Privacy standards
3. Security standards protecting confidentiality and integrity of health information
4. Unique health identifiers for individuals, employers, health plans and clinicians

Standardization of Electronic Data. Proposed data standards for electronic health care claims published in the Federal Register Sept 23 2005:

- Claims, eligibility, enrollment, etc. will use ANSI X12 N
- Pharmacy transactions will use NCPDP
- Diagnoses will use ICD-9
- Procedures will use ICD-9
- Physician services will use CPT codes
- Dental services will use CDT codes
- Lab services and other clinical services will use LOINC
- Communication will be based on HL7 standards using XML programming language
- Previously more than 400 EDI (electronic data interchange) formats are used
- Ironically, paper claims can still exist but when they are submitted electronically they must comply [39]

Privacy Rules

- HIPAA covers all health information and records regardless, whether they are electronic, oral or written
- Clinicians need not obtain patient permission (consent) for use of health information for treatment or payment. There has been confusion regarding this particularly in an emergency situation [40]
- HIPAA requirements include
  - internal protection of medical records
  - a written statement of privacy practices to patients
  - employee privacy training
  - privacy complaints must be addressed
  - designation of a privacy officer
  - de-identifiable patient information, when possible
- Enforcement will be by the DHHS Office for Civil Rights (OCR). The OCR will:
  - assist voluntary compliance efforts
  - respond to questions
  - investigate complaints
  - respond to state requests for exceptions
  - conduct compliance surveys
  - Assess monetary penalties or criminal prosecution

- All must comply
- Penalties from $100-$250 K and imprisonment for up to 10 years possible
- JCAHO and Medicare inspections will likely look at HIPAA compliance [41]

Security Standards

- There are 18 security standards. Implementation specifications are either required or addressable. If addressable, the practice must document why it can not be implemented, otherwise it is required
- Standards are divided into: administrative, physical and technical:
  - o **Administrative:** policies and procedures that implement security measures to protect health information and manage the conduct of employees
  - o **Physical:** policies and procedures that protect practice health information from natural disasters and intrusion. Example, locking charts up at the end of the day
  - o **Technical:** policies and procedures that control access to protected health information such as secure log-ons and authentication

Unique Identifiers

- HIPAA requires the use of a unique identifier for clinicians, health plans, employers and patients
- The National Committee on Vital and Health Statistics will assist DHHS
- Date to begin is May 2005 and compliance by May 2007
- National Provider System is part of DHHS and will issue a 10 digit numeric identifier or National Provider Identifier to clinicians
  It will replace, as an example, the UPIN and insurance provider IDs. Deadline is May 2007 but may be one year later for small physician groups
- National Employer Identifier will use the employer identification number (EIN)
- National Health Plan Identifier is on hold
- National Patient identifier is on hold [42]

As an example of the utility of identifiers: a patient enters the healthcare system and provides the hospital or office with identification using a smart card. The information is then exchanged electronically with the insurance company (payer), resulting in faster payments.

Benefits of HIPAA

- Patients can ask to see, copy and amend their medical records
- Patients can request an audit of who has accessed their record for the past six years
- States can opt for more stringent privacy than HIPAA, but not less

Limitations of HIPAA
- No consent needed for information sent to insurance companies
- Sensitive past medical information could become public
- Health information can be released to businesses looking to recall or replace drugs or devices
- Patients can not sue for breach of privacy under HIPAA
- Business associates of a covered entity can receive your health information without your consent
- Law enforcement may have limited access to protected health information (PHI) without consent
- There are no HIPAA police; system is complaint driven only. Of 18,000 complaints submitted, only two have been prosecuted [43]
- Will malpractice insurance cover a HIPAA violation?
- HIPAA needs to be considered with any new technology purchase such as a PDA, wireless system or RHIO if patient data will be transmitted. Process needs to be password protected and encrypted
- Will HIPAA slow down necessary innovation and modernization?
- Will the Privacy Rule impede research in general? [44]
- Those not covered by HIPAA:
  - Life insurance companies
  - RHIOs
  - Worker's compensation
  - Internet self help sites
  - Law enforcement agencies
  - Researchers who get information from primary care managers
  - Healthcare screening
  - Agencies that deliver social security and welfare benefits [45-46]

Conclusion

Interoperability makes the world go around. Without it technology would not work the same in each state. Data standards are mandatory for electronic health records but are still a work in progress. Until that happens we will struggle to build regional health organizations. RHIOs will also be impeded by an almost universal lack of a business model for sustainment. HIPAA adds privacy and security standards but also creates new hurdles to overcome.

References

1. Commission on Systematic Interoperability www.endingthedocumentgame.gov (Accessed January 10th 2006)
2. Weier S Commission releases 14 Interoperability Recommendations. www.ihealthbeat.org October 26 2005. (Accessed October 27 2005)
3. Certification Commission for Healthcare Information Technology www.cchit.org (Accessed January 15 2006)
4. Health IT World http://www.health-itworld.com/newsitems/2006/may/05-25-06-tepr

5. Halamka JD. Harmonizing Healthcare Data Standards. J Healthcare Info Man 2006;20:11-13
6. HITSP delivers interoperability specifications. HIMSS Insider November 2006; 21
7. Executive Order: Incentives for the Use of Health Information Technology and Establishing the Position of the National Health Information Technology Coordinator http://www.whitehouse.gov/news/releases/2004/04/20040427-4.html (Accessed February 18 2006)
8. Summary of Nationwide Health Information Network (NHIN). Request for Information (RFI) Responses June 2005 www.hhs.gov/healthit/rfisummaryreport.pdf (Accessed January 30 2006)
9. Office of the National Coordinator for Health Information Technology http://www.dhhs.gov/healthit/   (Accessed October 10 2005)
10. Scalese D. Which way RHIO? Hospitals and Health Network http://www.hhnmag.com/hhnmag_app/jsp/articledisplay.jsp?dcrpath=HHNMAG/PubsNewsArticle/data/2006June/0606HHN_InBox_Technology2&domain=HHNMAG (Accessed March 3 2007)
11. HealthAlliant http://www.healthalliant.org/ (Accessed February 18, 2006)
12. Center for Information Technology Leadership http://www.citl.org/ (Accessed February 18 2006)
13. Sundwall D. RHIO in Utah, UHIN HIMSS Conference June 6 2005
14. What Killed the Santa Barbara County Care Data Exchange? March 12 2007. www.ihealthbeat.org (Accessed March 13 2007)
15. Root J. Utah Health Information Network AHRQ First Annual Meeting June 8 2005
16. Egan C. Kentucky Electronic Health Network could serve as national model www.ihealthbeat.org (Accessed March 10 2005)
17. Indiana Health Information Exchange http://www.ihie.com/default.htm (Accessed February 18 2006)
18. HealthAlliant http://www.healthalliant.org/ (Accessed February 18 2006)
19. HealthVision www.healthvision.com (Accessed February 18, 2006)
20. Agency for Health Research and Quality www.ahrq.gov (Accessed February 22 2005)
21. Tennessee Blues, Cerner view shared project as unique HealthITNews. www.healthcareitnews.com May 23 2005 (Accessed May 24 2005)
22. IOM. Patient Safety: Achieving a new standard of care. 2004. http://www.nap.edu/books/0309090776/html/ (Accessed February 22 2005)
23. Kim, K Clinical Data Standards in Health care: Five case studies. California Healthcare Foundation July 2005 www.chcf.org (Accessed January 1 2006)
24. Health Level Seven HL7 www.hl7.org (Accessed December 20 2005)
25. DICOM http://medical.nema.org (Accessed December 21 2005)
26. Institute of Electrical and Electronics Engineers www.ieee.org (Accessed December 21 2005)
27. LOINC www.loinc.org (Accessed December 23 2005)
28. ELINCS. http://www.chcf.org/topics/chronicdisease/index.cfm?itemID=108868 (Accessed November 24 2006)
29. RxNorm http://www.nlm.nih.gov/research/umls/rxnorm_main.html (Accessed December 23 2005)
30. NCPDP www.ncpdp.org (Accessed December 22 2005)
31. SNOWMED-CT www.snowmed.org (Accessed December 23 2005)
32. Joch A, A blanket of SNOMED Federal Computer Week Nov 14 2005;s46-47
33. Elkin PL et al Evaluation of the content coverage of SNOMED-CT: ability of SNOMED Clinical terms to represent clinical problem lists Mayo Clin Proc 2006;81:741-748
34. ICD-9 www.who.int/whosis/icd10/othercla.htm (Accessed December 21 2005)
35. Featherly K ICD-9-CM: An uphill struggle Healthcare Informatics Oct 2004: 14-16
36. Weier S. Letter Encourages Congress to promote ICD-10. www.ihealthbeat.org May 19 2006 (Accessed February 22 2005)
37. CPT www.ama-assn.org/ama/pub/category/3113.thml (Accessed December 20 2005)
38. National Consumer Health Privacy Survey 2005 www.chcf.org (Accessed December 28 2005)

39. Federal Register Vol. 74 No 184 September 23 2005
http://a257.g.akamaitech.net/7/257/2422/01jan20051800/edocket.access.gpo.gov/2005/pdf/05-18927.pdf (Accessed October 1 2005)
40. Touchet http://ps.psychiatryonline.org/cgi/content/full/55/5/575
(Accessed December 20 2005)
41. A HIPAA Primer http://www.hipaadvisory.com/regs/HIPAAprimer.htm (Accessed December 19 2005)
42. Use of the National Unique Identifiers
www.ehcca.com/presentations/HIPAAWest2/barry.ppt (Accessed December 20 2005)
43. Hayes HB Government Health IT June 2006 pp28-33 www.govhealthit.com (Accessed July 15 2006)
44. Wilson JF Health Insurance Portability and Accountability Act Privacy Rule Causes Ongoing Concerns among Clinicians and Researchers Annals of Int. Med. 2006;145:313-316
45. Centers for Medicare and Medicaid Services. HIPAA.
http://www.cms.hhs.gov/HIPAAGenInfo/ (Accessed February 18 2006)
46. Privacy Rights Clearing House. www.privacyrights.org (Accessed February 18 2006)

 # Chapter 4: Patient Informatics

Learning Objectives

After reading this chapter the reader should be able to:
- Identify the origin of patient informatics and the role of the Internet in patient education
- List the standard features of a patient web portal
- Compare and contrast the various types of personal health records and their projected benefits
- Describe the evolving role of secure e-mail and virtual visits as a new form of patient-physician communication

Introduction

Patient Informatics is a new aspect of Medical Informatics that largely reflects the empowered healthcare consumer. Patients are aware that many non-healthcare businesses are automating and modernizing their business processes to attract a larger market share. ATM machines, as an example, can provide cash in a few minutes regardless of where you are located worldwide. This innovation required re-engineering and the acceptance of universal standards, not unlike many aspects of information technology.

In this chapter we will discuss four aspects of patient informatics:
1. The Internet for patient medical education
2. Web portals for patient access to their healthcare systems
3. Personal health records
4. Patient-physician communication via e-mail and e-visits

These forms of patient informatics have occurred in the past five to ten years. Since little has been written in the medical literature about this topic, we must rely primarily on surveys and expert opinions. A 2006 Harris Interactive survey of 2,624 adults determined the desired informatic features (Table 4.1).[1]

Desired informatics features	% Respondents
E-mail reminder of upcoming appointments	77%
Schedule medical appointments online	75%
Communicate with physician via e-mail	74%
Receive test results via e-mail	67%
Access to their electronic health record	64%
Home monitoring that would transmit information to physician's office	57%

Table 4.1. Patient Informatics features desired

Note that most of their desires are easily achievable but unavailable in many healthcare systems.

The Internet for patient medical education

Multiple surveys have confirmed that the Internet is the premier medical resource for both patients and medical professionals. This has occurred in the incredibly short time span of just one decade. According to a 2005 survey by the Pew Internet & American Life Project: 87% of respondents aged 29-40 use the Internet, compared to only 21% aged 70 or older. In those who use the Internet, all age groups search for health information. Furthermore, approximately 63% of adults have broadband access at home or at work. The individuals most likely to seek health related information online are women, age younger than 65, college graduates, those with more online experience and those with broadband access.[2] It should be pointed out that this is United States data as many countries are more "wired" and many more are less connected (Somalia .002%).[3] Additionally, there is a significant digital divide between ethnic and racial groups in the United States. In a 2003 study by the National Center for Education Statistics, 54% of white students used the Internet at home as compared to about half that for Hispanic and black youth.[4]

In another survey by Opinion Research Corporation it was noted that the top healthcare worries prompting Internet searches were: cost (41%), quality (25%), medical errors (16%) and all others (18%).[5] Table 4.2 shows the types of medical topics searched on the Internet. Although specific diseases were most commonly researched it was of interest that insurance issues, hospitals and doctors were also commonly researched.[2]

Healthcare Topics Searched	% Users Who Searched
Specific Disease	63
Medical Treatment	47
Diet and Nutrition	44
Exercise	36
Medication Issues	34
Alternative Medications	28
Insurance Companies	25
Depression	21
Doctor or Hospital	21

Table 4.2. Healthcare topics searched

Recent events confirm that patients are becoming more discriminating in their choices of all aspects of healthcare. No longer do they automatically accept the opinion of their physicians. In a Harris poll it was shown that 57% of patients discussed their Internet search with their physician and 52% searched the Internet after talking to their physician. Eighty nine percent felt their search was successful suggesting confidence in the Internet as the new health library.[6] Excellent medical web sites exist but searches may yield low quality, non-evidence based answers, particularly when personal web sites are searched. In one study of Internet searching for the treatment of childhood diarrhea, 20% of searches failed to match the guidelines published by the American Academy of Pediatrics. [7]

Multiple reasons have been suggested for the increased use of the Internet in the healthcare arena:

- Healthcare is becoming more patient-centered in general. This has been promoted by the Institute of Medicine in all of their publications [8]
- Quality patient education web sites abound
- Most non-medical businesses are offering an online method to promote services or make contact
- The increasing use of medical blogs, podcasts and wikis as part of the Web 2.0 movement [9]
- Patients are becoming impatient about finding the right answers, the best physicians, the best hospitals and the best medical care at the lowest cost
- Baby Boomers are well educated and have more disposable income. This results in higher expectations. There is some evidence that patients expect their physicians to be tech savvy and would consider switching if they are not
- Healthcare organizations are using information technology as a marketing tool to attract patients
- Computers are ubiquitous, as are broadband connections. Searching the Internet is now very fast and easy
- Patients receive less face time with their physicians causing some patients to turn to the Internet for answers

Examples of Online Patient Education web sites

The following are only a sample of the many valuable patient education web site available today.

WebMD. With more than 30 million people visiting this site monthly, it should be considered one of the true standard bearers. They have an extensive health library with top topics also listed for men, women and children. Treatment and drug information is available as it medical news. A symptom checker tool provides a patient with a simple differential diagnosis of what might be wrong with them based on their symptoms, age and gender. A daily e-mail newsletter is offered that can be customized to a patient's concerns. The only negative is some commercial influence. [10]

Revolution Health. This site was partially launched in late 2006 with plans to be full service in the spring of 2007. It is backed by America Online as well as personalities such as Colin Powell and Carly Fiorina. Personal health records are an option, as are disease information, forums, health calculators, physician finder and symptom checker. Members can rate physicians and hospitals, in addition to treatments and medications. The also offer an insurance marketplace to discuss and compare insurance options. Although these services are free, they also offer a fee-based membership that will allow a member to call to discuss a health condition or need 24 hours a day. [11]

MedlinePlus. The premier free patient education site developed by the National Library of Medicine and the National Institute of Health that links to the best and most respected web sites, such as the Mayo Clinic. MedlinePlus is ranked as the top information/news web site on the American Customer Satisfaction Index of Federal government web sites.[12] In spite of its high marks, many patients and clinicians do not know about this site and many healthcare organizations pay for patient education that could be obtained for free. Features of the web site include:

- 700+ health topics available in English and Spanish
- Search results with high quality references (Figure 4.1) [13]
- Drug information
- Health encyclopedia and dictionary
- 165 tutorials
- Videos of surgical procedures
- Directories to locate physicians and hospitals
- Clinical Trials.gov to determine where certain diseases are being studied
- Other health databases
- Health news

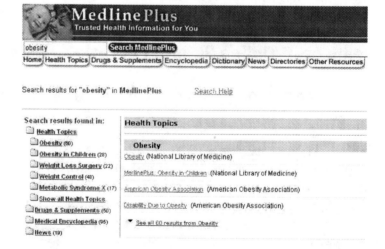

Figure 4.1 Search results for obesity (courtesy MedlinePlus)

Healthfinder. Developed by the National Health Information Center and the Department of Health and Human Services to provide general information for the average patient. It is very similar to MedlinePlus.[14]

Florida State University Medical Library. Developed by Nancy Clark, this site includes MedlinePlus, Healthfinder and twenty two other general patient education web sites. Importantly, it links to twelve Pediatric and fourteen specialty resources. It would likely cover any subject of interest to the average patient.[15]

Healthwise. Multiple companies sell patient education for use on commercial medical web sites. Healthwise is a not-for-profit company that provides more than 6,000 medical topics in their knowledgebase. Other features include decision making tools, "take action tools"
for chronic diseases and over 1,000 illustrations. [16]

Patient web portals

Web portals are web based programs that patients access for health related services. A web portal can be a standalone program or it can be integrated with an electronic health record. Patient portals began as a web based entrance to a healthcare system for purposes of learning about a hospital, healthcare system or physician's practice. Also, they were clearly a marketing ploy to attract patients who were Internet savvy. Currently, patient portals offer multiple patient services: online registration, medication refills, lab results, electronic visits, patient education, storage of personal health records, appointments, referrals, secure messaging, bill paying, document uploading and patient logs for vital signs and tracking of other health data.
Figure 4.2 shows other ways patients can communicate with their physician besides face to face. Time will tell how well these new forms of e-communication work, how cost effective they are, how well they are reimbursed and how well they are liked by physicians and patients.

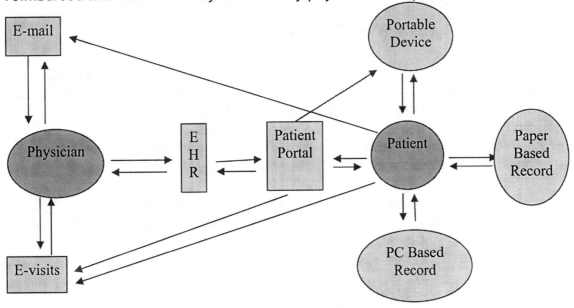

Figure 4.2. Patient, physician and chart interactions

A minority of web portals actually integrate with an EHR, which means that most patient data has to be manually inputted. In the future when EHRs become more widespread, selected patient lab results will automatically post to the patient portal, thus saving time and money. Patients will be able to also access parts of their electronic records. A 2006 Harris Interactive study showed that 83% of patients wanted lab tests online and 69% wanted online charts to manage chronic conditions.[17] Although several studies have shown interest in having access to lab results it remains to be seen if that would change behavior or outcomes. Are these patients primarily college educated and tech savvy? Do they desire results because physician's offices are slow to provide results? In a study at Beth Israel hospital, patients who accessed their portal PatientSite were younger and with fewer medical problems. They tended to access lab and x-ray results and use secure messaging more than a non-enrollee group.[18] In another survey by Connecting for Health, patients were asked "I think that having my health information online would" with the following responses:

- 71% said it would clarify doctor instructions
- 65% said it would prevent medical mistakes
- 60% said it would change the way I manage my health
- 54% said it would improve the quality of care [19]

Little is written about the benefits of patient web portals. McLeod Health System in Florence, South Carolina used online scheduling as part of the portal NexSched and was able to demonstrate fewer "no shows" and fewer claims denials. They predict a savings of about $1 million dollars yearly as a result of this program.[20] Patient portals and the other new forms of e-communication have gained the attention of several organizations. Robert Wood Johnson Foundation has awarded $2.45 million dollars to six organizations to study the effect of patient portals on chronic disease.[21]

Examples of Patient Web Portals

MySaintAls. This is the portal for Saint Alphonsus Medical Center in Boise, Idaho. This comprehensive portal offers all of the standard features as well as the unique Patient Vault. They charge $10 monthly to upload (scan) and store patient records on their server. Lab results are accompanied by a separate program that explains the significance of the results and likely reduces the number of routine questions. [22]

Epic MyChart. A patient portal integrated with a well established electronic health record system. Over 50,000 patients have enrolled in MyChart at Palo Alto Medical Foundation as of November 2005. An interactive demo is available on the web site. [23-25]

RelayHealth This portal has traditionally been a standalone product but in 2006 they offered integration with EHRs such as Allscripts, McKesson and NexGen. In

2006 they were acquired by McKesson Corporation. Their business model is to charge physicians $49 monthly to cover e-visits using their secure messaging site. They offer the features of most current portals and in addition offer structured questions for e-visits. A demo is also available on their site. [26]

ReachMyDoctor. This site is aimed at improving communication with the doctor's office and offers two options:
- Free: schedule appointments, request medication refills, request a referral and address billing and insurance issues
- Subscription: for $8.95 monthly a patient can ask the physician non-urgent questions via secure e-mail. Physicians must be part of the network [27]

My HealtheVet. A portal that integrates with the Veterans Affairs' EHR and offers lab results, appointments, a personal health record (PHR), medication refills, patient education and online monitoring of: activity, food intake, oximetry, blood pressure, glucose and weight. Key components will be available by December 2007. A pilot project is underway (eVault) to provide select parts of the EHR to the patient. [28]

Personal Health Records

According to the American Health Information Management Association the personal health record (PHR) is:
> "an electronic, universally available, lifelong resource of health information needed by individuals to make health decisions" [29]

The Institute of Medicine promotes PHRs by stating patients "should have unfettered access to their own medical information". [30] The first principle endorsed by the Personal Health Technology council is that "individuals should be able to access their health and medical data conveniently and affordably". [31]

In April 2004 President Bush created the National Health Information Network. The vision of the NHIN is largely based on a universal electronic health records system that would become interoperable within the next decade. Additionally, a patient portal would be part of the electronic health record allowing patients to upload or download personal health information. Most experts feel this is the ultimate method of providing personal health records but believe that it may be ten or more years in the future. Currently, fewer than 15% of all outpatient practices have an electronic health record and fewer than 2% can generate a personal health record. [32] What is needed therefore is an interim solution so medical information can be available to the patient and physician anytime and anywhere. Nowhere is it more important than in the military with a high turnover of both patients and physicians. In the case of deployment there is an acute need for personal health records, as none exist in the battlefield currently. The US Army has recognized this problem and developed a software program called the Battlefield Medical Information System-Tactical (BMIST) that utilizes flash memory for its "personal information carrier" (electronic dog tag).[33] Although the Department of Defense (DOD) has an electronic health record (AHLTA) it can not generate a personal health record. Similarly, many patients are seen both in the (DOD) and the Veterans Affairs (VA) system but to date these systems do not universally communicate and will not in the foreseeable future.[34]

Interest in PHRs comes from multiple sources. In 2002 the Markle Foundation established *Connecting for Health*, a public-private collaboration to promote better information sharing between doctors and patients. In their July 2004 position paper they suggested: PHR development should be increased, PHRs will educate patients about their health and common data standards are a logical starting point. In one of their surveys 61% of respondents agreed that they should have access to their medical information "anytime, any place".[35] A 2004 Harris Interactive survey of over 2,000 adults demonstrated that 42% kept personal or family health records but only 13% stored their records electronically.[36] Another survey conducted in 2004 by Bearing Point noted that over 50% of respondents would be interested in carrying their medical records in a portable device, accessible in an emergency.[37] The Centers for Medicare and Medicaid Services released a "Request for Information" (RFI) about PHRs in July 2005 to determine its future direction.[38] They are well aware of the need for better patient information sharing and storage in older patients who are on multiple medications and have multiple physicians. Most recently, attention has been given to electronic health records and personal health records after hurricanes Katrina and Rita. As a result, Blue Cross Blue Shield of Texas created personal health record like summaries of care from insurance claims data, available to both physicians and patients.[39]

In late 2006 America's Health Insurance Plans (AHIPs) and Blue Cross-Blue Shield Association announced a comprehensive plan to supply PHRs to their members by 2008. Importantly, they have established core data standards and much of the information will come from claims and administrative data. With established standards, PHRs can be shared between different insurance companies, should a patient move or change coverage. [40-41] Aetna will also deploy PHRs for its members in the near future. Of interest, they will use *CareEngine*, a software rules engine that reviews the PHRs and gives personalized alerts (called care considerations) to patients and physicians about how to improve medical care.[42]

Also appearing in late 2006 was a PHR system (Dossia) that was founded by Applied Materials, BP America, Intel, Pitney Bowes and Walmart. Data will derive from insurers, pharmacies and physicians. Information will be web based and able to be downloaded to portable media.[43]

Personal Health Record Formats. There are more than fifty personal health record products on the market currently.[44] These products are available in multiple formats:

1. *Paper based*. Free forms can be downloaded from the Internet and data entered by the patient

2. *Personal computer (PC) based*. A software program creates a PHR and stores it on the PC but this information can not be shared with others. Hybrids exist that allow uploading of the information to a web site.

3. *Web based*. Most are commercial sites that are secure and can be accessed from a distant site. A minority of PHRs reside in patient portals that connect to an electronic health record system.

4. *Mobile technology*. Patient information can be downloaded to secure digital cards, USB drives, "Smart Phones", BlackBerries and "Smart Cards". Most USB programs synchronize to a web based portal where patient information is also stored. Mobile technology offers several unique advantages. It is not dependent on the Internet for operation and is truly portable. The Internet may not be available in the following situations: deployed military; young children and the elderly who may not use it and during national disasters. Another argument for portability is the fact that on average 41 million Americans move each year.[45]

Flash memory devices (USB drives and secure digital cards) are extremely popular based on the fact that the technology is "plug and play", is very fast, has extensive memory (up to 4 gigabytes) and is relatively inexpensive. Recent studies have shown tremendous growth in the use of USB drive devices and to a lesser degree secure digital memory cards.[46] USB drive memory devices can be used on virtually any computer produced since 2000. A more rugged waterproof USB drive is available with 32MB-1GB of memory. Secure digital cards have the added benefit of being able to be used in PDAs, "smart phones", computers and cameras. In 2005, a combination SD/USB drive was developed that is available with either 512 MB or 1 GB of memory.[47] In 2006 a new USB drive in the form and size of a credit card with memory of 128-2000 MB [48] appeared and is now available as a PHR.[49]

It is unknown how the average patient will want to store or carry their mobile PHRs. Given the soaring popularity and expanding features of smart phones they may become the mobile storage of choice for PHRs in the future.[50]

Smart cards generally have less than 64 K of memory, although this seems to be improving. They require unique scanners that are expensive and not available universally. Most cards are used for patient authentication in the healthcare environment.

Six key questions remain unanswered about PHRs:

1. What medical information should be in the personal health record?

2. Who should input the information; the patient, the family, the clinician or the healthcare organization?

3. Should the information be downloaded manually or should it be downloaded via an interface with the hospital or electronic health record system?
4. Will mobile PHRs prove to be more valuable than PHRs stored on a web site or personal computer?
5. Who will pay for the data download, the storage device and the maintenance of a PHR system?
6. Do we currently have enough security systems in place to protect personal data and comply with the Health Insurance Portability and Accountability Act (HIPAA)? Are we sure that mobile PHR models would be secure? A 2007 article pointed out that it was easy to hack into a computer's data by using a USB device. [51]

Although there are no standards for personal health records, many experts feel that the most logical PHR data to download would be the same as used in the "Continuity of Care Record" (CCR). The CCR has the essential information one would need for an emergency or a hospital discharge. The administrative and medical information could be web based or stored on a portable memory device and relayed to another physician or hospital. [52-54]

Thus far, personal health records have been voluntary, placing the burden of downloading and maintaining health information on the patient. A busy physician's office is not likely to want this additional responsibility without reimbursement. In hospitals or clinics with electronic health records or available claims data, software programs could be written to automatically download patient data to portable storage devices. Although this theoretical method has the widest appeal it is by far the most expensive and raises serious security issues. As PHRs develop more user friendly features, perhaps the appeal to the average consumer will increase. Some PHRs, for example, will provide alerts such as medications about to expire or upcoming medical appointments. An ideal business model for personal health records does not exist. Some studies suggest the patient is willing to purchase their own PHR if the price is low and others suggest insurance companies are the most likely to play a major role. Theft of personal health information is a definite concern, but with more sophisticated encryption and authentication it is likely this will be not be a major obstacle.

Secure patient-physician e-mail and e-visits

The vast majority of Americans today use e-mail, however few physicians routinely use e-mail to communicate with their patients. [55] According to a 2005 survey by Harris Interactive, 80% of patients who go online would like to communicate with their physicians. [56] On the other hand, in another survey in 2005 by the Center for Studying Health System Change, only 24% of all physicians used e-mail to communicate with patients, [57] compared to 3.4% reported in a

*Digital Rx:
Take two aspirin
and e-mail me in
the morning*

*New York Times
3/2/2005*

2004 study.[58] Furthermore, 41% of respondents in a 2001 Harris Interactive study, found it very frustrating to see a physician in person when they thought a telephone call or e-mail would suffice.[59] Multiple studies suggest patients want to communicate with their physicians via e-mail but that enthusiasm is not shared by physicians.[60] Physicians cited the following reasons for not using e-mail: liability (56%), poor compensation (45%), privacy issues (43%), staff not trained (30%) and the feeling that face to face visits are better (27%).[61] E-visits are electronic visits using secure messaging and not routine e-mail. This has the advantage of much better security and privacy and the ability to have a third party involved in the billing structure. Patients and physicians must utilize a username and password to log onto a secure web site in order to conduct an e-visit. Numerous vendors such as RelayHealth and MedFusion exist to provide a venue for e-visits in addition to their patient portal features. Patients are seeking better service by wanting to use e-mail but their expectations might be unrealistic. In a survey of 950 primary care patients in Texas, 62% of respondents expected the results of their lab results by e-mail in less than 24 hours. [62]

Multiple benefits of e-mail communication have been pointed out. The communication is asynchronous so physicians can answer at their convenience and avoid "phone tag". Overhead is less for electronic messages compared to phone messages and they are self-documenting. Patients tend to lose less time from work. Potential disadvantages might include: indigent patients less likely to use service, inability to examine patient, potential for communication error, possible slow responses, security issues and the potential to be overwhelmed.[63-64] Thus far, the concerns have not been borne out by published studies.

Some authorities feel the e-visits have a bright future. A Price Waterhouse study estimated that 20% of outpatient visits could be eliminated by using e-visits.[65] A new CPT code 0074T has been developed for e-visits. [66] Nevertheless, e-mail conduct guidelines have to be established to prevent abuse.[67] Guidelines also need to be established to define what constitutes an e-visit in order for insurance companies to reimburse the electronic visit.

Several reports suggest that e-mail and e-visits don't negatively impact a physician's productivity.[68-70] The consensus is that minor complaints can be dealt with more efficiently electronically, thereby allowing sicker patients to be seen in person. It has also been pointed out that if the patient provides a history during the e-visit and still has to be seen face to face, the physician has the advantage of knowing why the patient is there, therefore saving time.

In spite of the enthusiasm for e-mailing physicians, most patients are not willing to pay more than minimal co-pays for an e-visit.[71] In a study by RelayHealth they were able to demonstrate that, compared to controls, the patients who had e-visits had lower insurance claims. The profit more than paid for the $25 physician charge and the $0-10 patient co-pay. Importantly, 50%

were less likely to miss work and 77% said it only took 10 minutes for the e-visit. Patient and physician satisfaction levels were good.[72] Pilot projects and studies are underway to evaluate e-visits. BlueCross/Blue Shield of Tennessee and other regions are reimbursing physicians for electronic visits.[73-76] The University of California at Davis Health System has been performing e-visits since 2001 and states that 80% of insurance companies in their region support the concept. Participating physicians seem to be more cost effective and physicians are reaping a bonus.[77] An excellent review of Patient-Provider communication can be found in a 2003 monograph by the First Consulting Group.[78]

Conclusion

Many patients have cast their vote in favor of a more user-friendly healthcare system. They desire rapid access to medical answers via the Internet and rapid access to their healthcare system through their web portal. They would like to have storage of their personal health records somewhere but are reluctant to pay for it. Lastly, they want to communicate with their clinicians via secure messaging and if necessary initiate an e-visit. It will take time to see if patients, clinicians and payers align to make this a reality.

References

1. PR Newswire. http://sev.prnewswire.com/health-care-hospitals/20060922/NYF08222092006-1.html (Accessed October 26 2006)
2. The Patient and the Internet. Pew Internet & American Life Project. May 2005 www.pewinternet.org   (Accessed January 5 2006)
3. Matthew Zook. Zooknic http://www.zooknic.com/ (Accessed September 23 2006)
4. Hayes, David Midday Business Report. The Kansas City Star September 5 2006 (Accessed 20 September 2006)
5. Opinion Research Corporation—Siemens. Survey 2003. www.Informationtherapy.org/conf_mat05/hallppt.pdf (Accessed January 5 2006)
6. Harris Poll http://harrisinteractive.com/harris_poll/index.asp?PID=584 (Accessed January 10 2006)
7. McClung H The Internet as a source for current patient information. Pediatrics 1998;101: p. e2
8. Institute of Medicine www.iom.edu (Accessed December 9 2006)
9. How Web 2.0 is changing medicine www.bmj.com/cgi/content/full/333/7582/1283 (Accessed March 15 2007)
10. WebMD www.webmd.com (Accessed March 15 2007)
11. Revolution Health www.revolutionhealth (Accessed March 15 2007)
12. American Customer Satisfaction Index of Fed Govt web sites December 2004 survey. (Accessed January 5 2005)
13. MedlinePlus www.medlineplus.com (Accessed January 12 2007)
14. Healthfinder www.healthfinder.gov (Accessed January 12 2006)
15. Florida State University Medical Library Patient Education http://www.med.fsu.edu/library/PatientEd.asp (Accessed February 3 2006)
16. Healthwise www.healthwise.org (Accessed March 4 2007)
17. First Health, HarrisInteractive. Consumer Benefits Health Survey. Executive Summary (Accessed January 12 2006)

18. Weingart SN et al Who Uses the Patient Internet Portal? The PatientSite Experience JAMIA 2006;13:91-95
19. Connecting for Health. Markle Foundation June 2003. www.connectingforhealth.org. (Accessed February 10 2006)
20. Egan C. Online Patient Scheduling Improves Time, Cost Efficiency www.ihealthbeat.org October 28 2004 (Accessed December 1 2004)
21. Broder C. Foundation Grants Fund Patient Web Portal, Disease Management Initiatives www.ihealthbeat.org October 4 2004 (Accessed December 1 2004)
22. Alphonsus Medical Center. www.MySaintAls.com (Accessed October 2 2005)
23. Epic. http://www.epicsystems.com/Software/eHealth.php#MyChart (Accessed February 28 2006)
24. Palo Alto Medical Foundation www.pamfonline.org (Accessed December 3 2005)
25. Palo Alto Medical Foundation enrolls 50,000 patients to online service. November 11 2005 www.ihealthbeat.org (Accessed January 10 2006)
26. RelayHealth www.relayhealth.com (Accessed October 4 2005)
27. ReachMyDoctor www.reachmydoctor.com (Accessed January 20 2007)
28. My HealtheVet http://www.myhealth.va.gov/ (Accessed February 28 2006)
29. American Health Information Management Association www.ahima.org (Accessed July 5 2006)
30. Crossing the Quality Chasm: A New Health System for the 21st Century Institute of Medicine 2001 The National Academies Press p. 8
31. Personal Health Technology Council www.markle.org (Accessed September 1 2006)
32. Medical Group Management Association and University of Minnesota School of Public Health Study September 2005 http://www.ahrq.gov/news/press/pr2005/lowehrpr.htm (Accessed September 20 2005)
33. Telemedicine and Advanced Technology Research Center www.tatrc.org (Accessed October 1 2005)
34. ihealthbeat http://www.ihealthbeat.org/index.cfm?Action=dspItem&itemID=100265 November 20 2003 (Accessed October 1 2005)
35. Connecting for Health. Working group on policies for sharing information between doctors and patients. July 2004 http://www.connectingforhealth.org/resources/wg_eis_final_report_0704.pdf (Accessed October 1 2005)
36. HarrisInteractive market research http://www.harrisinteractive.com/news/newsletters/healthnews/HI_HealthCareNews2004Vol4_Iss13.pdf (Accessed October 1 2005)
37. Press Release SanDisk www.sandisk.com/pressrelease/20050214b.htm (Accessed October 5 2005)
38. Centers for Medicare and Medicaid Services www.cms.hhs.gov and http://www.gcn.com/vol1_no1/health_IT/36422-1.html (Accessed November 1 2005)
39. Insurer makes electronic patient summaries available for hurricane evacuees http://www.healthcareitnews.com/NewsArticleView.aspx?ContentID=3710 (Accessed December 1 2005)
40. Insurers to Provide Portable, Interoperable PHRs. www.ihealthbeat.org December 14 2006. (Accessed December 14 2006)
41. Industry Leaders Announce Personal Health Record Model; Collaborate with Consumers to Speed Adoption. http://bcbshealthissues.com December 13 2006 (Accessed December 14 2006)
42. Aetna Introduces Powerful, Interactive Personal Health Record. www.aetna.com/news/2006/pr_20061003.htm. (Accessed December 14 2006)
43. Providers Want More Details on Employer PHR Initiative. www.ihealthbeat.org December 12 2006)

44. Personal Health Records in the Marketplace http://library.ahima.org/xpedio/groups/public/documents/ahima/pub_bok1_027459.html (October 20 2005)
45. Census data. http://www.census.gov/population/pop-profile/p23-189.pdf (Accessed October 10 2005)
46. Gartner Dataquest http://www.gartner.com/press_releases/pr15may2003a.html (Accessed October 10 2005)
47. SanDisk Corporation www.sandisk.com (Accessed October 10 2005)
48. Worldwide smart phone market soars in Q3 http://www.geekzone.co.nz/content.asp?ContentId=5390 (Accessed January 2 2006)
49. Walletex www.walletex.com (Accessed August 10 2006)
50. My Medpass www.mymedpass.net (Accessed August 10 2006)
51. Wright A, Sittig DF Security Threat Posed by USB-Based Personal Health Records Ann of Int Med 2007;146:314-5
52. The Continuity of Care Record www.findarticles.com/p/articles/mi_m3225/is_7_70/ai_n8570289/print (Accessed October 3 2005)
53. The Concept Paper of the CCR www.astm.org/COMMIT/E31_ConceptPaper.doc (Accessed October 5 2005)
54. PowerPoint presentation on CCR http://www.astm.org/COMMIT/3 (Accessed December 2 2005)
55. Slack WV A 67 Year Old Man Who e-mails his Physician JAMA 2004;292:2255-2261
56. Gullo K. Many Nationwide Believe in the Potential Benefits of Electronic Medical Records and are Interested in Online Communications with Physicians. HarrisInteractive Health Care Poll March 2005 www.harrisinteractive.com (Accessed October 10 2006)
57. Liebhaber A, Grossman J Physicians Slow to Adopt Patient E-mail. HSC September 2006 www.hschange.com. (Accessed October 10 2006)
58. Grant RW et al. Prevalence of Basic Technology Use by US Physicians. J Gen Int Med 2006;21:1150
59. Study Reveals Big Potential for the Internet to Improve Doctor-Patient Relations. Harris Interactive 2001 www.harrisinteractive.com (Accessed September 24 2006)
60. Medscape Poll December 2004 www.medscape.com (Accessed January 1 2005)
61. Physician Survey Jupiter Research June 2003 (Accessed October 2004)
62. 51.Couchman GR, Forjuoh MB, Rascoe TG E-mail communications in family practice: What do patients expect? J Fam Pract 2001;50:414-418
63. Car J, Sheikh A E-mail consultations in health care: 1-scope and effectiveness. BMJ 2004;329:435-438
64. The Changing Face of Ambulatory Medicine—reimbursing physicians for computer based care. American College of Physicians Medical Service Committee Policy Paper March 2003
65. Healthcast 2010:Smaller world, bigger expectations. Price Waterhouse Cooper. November 1999 www.pwc.com (Accessed February 3 2006)
66. Broder C What's in a code? www.ihealthbeat.org January 14 2004 (Accessed January 14 2004)
67. Kane B, Sands D Guidelines for the clinical use of electronic mail with patients" JAMIA 1998;5:104-11
68. Liederman EM Web Messaging: A new tool for patient-physician communication JAMIA 2003;10:260-270
69. Chen-Tan Lin An Internet Based patient-provider communication system: randomized controlled trial JMIR 2005.
70. Leong SL Enhancing doctor-patient communication using e-mail: a pilot study J Am Fam Pract 2005;18:180-8

71. Juniper Research October 2003 www.juniperresearch.com (Accessed December 10 2005)
72. The RelayHealth web visit study: Final Report www.relayhealth.com (Accessed January 2 2006)
73. Tennessee Hospital Pilots two e-mail programs www.ihealthbeat.org March 14 2005 (Accessed March 14 2005)
74. Blue Shield of California web communications pilot to enroll 1,000 physicians www.ihealthbeat.org April 20 2004 (Accessed April 20 2004)
75. Health Plan to pay Doctors for web visits www.ihealthbeat.org May 24 2004 (Accessed June 30 2004)
76. Microsoft Pilot Project to Test Online Physician Visits www.ihealthbeat.org January 10 2006 (Accessed February 2 2006)
77. UC Davis Virtual Care Study. Eric Liederman. Presented at AMDIS/HIMSS 2004
78. Online Patient-Provider Communication Tools: An Overview. First Consulting Goup. November 2003. California HealthCare Foundation www.chcf.org (Accessed September 20 2006)

Chapter 5: Online Medical Resources

Learning Objectives

After reading this chapter the reader should be able to:
- State the challenges of staying current for the average clinician
- Describe the characteristics of an ideal educational resource
- Describe the evolution from the classic textbook based library to the current online digital library
- Compare and contrast the different formats of digital libraries
- Identify the most commonly used free and commercial online libraries
- Describe the future of digital resources integrated with electronic health records

Introduction

> "Knowledge is Power"   Frances Bacon 1597

Trying to keep up with the latest developments in medicine is very difficult, primarily due to the accelerated publication of medical information and the significant time constraints placed on busy clinicians. It is likely that clinicians are in fact so busy that they have no idea what new educational resources are available to them. They would like to move from the "information jungle" to the "information highway" but who will show them the way? This chapter is devoted to those clinicians who are seeking rapid retrieval of high quality medical information.

What are the challenges clinicians face today?

- **Educational Challenge.** For example, 460,000 articles are added to Medline yearly.[1] The 2006 Physicians Desk Reference (PDR) is over thirty four hundred pages long making it exceedingly cumbersome to search for drug information.[2] It seems obvious that this is a disincentive to search for drug information and therefore is a patient safety issue. Standard medical textbooks are expensive and out of date shortly after publication. In addition, some argue that the descriptions of diseases are not always updated or evidence based.[3] Moreover, Shaneyfelt estimated that a General Internist would need to read 20 articles every day just to maintain present knowledge [4]

> "The complexity of modern American medicine is exceeding the capacity of the unaided human mind"
>
> **David Eddy**

- **Specialty to Primary Care Manager (PCM) Challenge.** Recommendations from specialty organizations take time to trickle down to the generalists. There is no standard way to disseminate information that is either reliable or particularly effective. National guidelines, usually written by specialists face the same challenges. Once there is a new standard of care for a disease such as diabetes, how do you get the word out, particularly to small or remote medical practices?

- **Translational Challenge.** Studies have shown that it may take up to ten or more years for research to be translated to the exam room (e.g.thrombolytics). [5] In a study by Antman, experts were also slow to make recommendations in textbooks even though high quality evidence was published many years prior.[6] On the other hand, many physicians are skeptical and wait for confirmatory studies. If they have been in practice for many years they may have witnessed the pendulum sweeping back and forth, for example, regarding the use of post menopausal estrogens. Recent studies often contradict older studies due in part to better study design and larger subject populations [7]

- **Evolutionary Challenge.** We can no longer teach "classic medicine" because diseases and their presentations change over time. This is best demonstrated by infectious diseases. Rocky Mountain Spotted Fever began to disappear as Lyme disease began to appear. Additionally, diseases were detected at more advanced stages in the older literature because lab tests were lacking, making clinical presentations more dramatic. Currently we tend to diagnose diseases earlier, before the patient has advanced signs and symptoms due to better tests. Medical resources therefore must reflect new evidence

- **Retention Challenge.** According to many studies there is an inverse relationship between current knowledge and the year of graduation from medical school. This was shown in a study by Ramsey that compared board scores of Internists and the number of years elapsed since certification [8]

How often do we actually have patient related questions and how often do we find answers?

- Covell reported that on average Internal Medicine physicians had two questions for every three patients seen and found the answers for only 30% [9]

- A study by Ely showed that Family Medicine physicians had 3.2 questions per 10 patients seen. The answer was pursued in only 36% of cases [10]

- In a primary care survey Gorman noted 56% of physicians pursued answers where they thought an answer existed and 50% of answers dealt with an urgent issue. Most physicians turned to other docs for answers and not the traditional medical library. Lack of time was the universal reason not to pursue answers in most studies.[11]

- In another study by Ely, the most common questions dealt with drugs, Ob-Gyn and adult infectious disease issues. Answers to 64% of questions were not pursued and physicians averaged less than two minutes per search. The most common resources used were books and colleagues and only two physicians performed literature searches.[12] It is important to point out that all of the above studies evaluated primary care physicians so the needs of other physicians such as surgeons are less clear. Also, after these studies were published software programs such as Epocrates appeared and significantly changed how we seek drug information

What is the state of medical libraries today?

In an excellent article by Lee entitled "Quiet in the Library" we see that traditional libraries are being forced to accommodate online resources or shut their doors. Many libraries are a great place to study because they are so quiet due to the fact that nobody uses them.[13-14] Florida State University Medical School is the newest US medical school and at this time offers 95% of its educational resources online (personal communication with Nancy Clark 2006).

How have we evolved from the traditional library to online resources?

Within a very short timeframe the Internet has become the educational resource of choice due to the speed of retrieval and depth of information. In a 2001 study by the American Medical Association it showed that 75% of physician practices had Internet access and 79% used it to research answers. Three out of ten practices had their own website.[15] These statistics continue to rise as does the availability of broadband access. The trend of physicians using the Internet as an online library has been closely mirrored by patients. The Internet now hosts more than 3 billion web sites. As an indication of growth, a Google search for the words "medical education" in 1995 by one of the authors yielded 760 results[16], whereas a search in 2006 yielded 49 million citations.

Medline became a reality in 1966 as it began archiving medical literature that was previously accessed manually through the *Index Medicus*. Today this National Library of Medicine resource contains over twelve million references from forty six hundred worldwide journals.[1] Using the Internet to access Medline is a logical progression but does it result in high quality answers quickly? Several studies have shown that finding an answer is difficult and takes too much time for a busy clinician.[17-18] Although a pertinent abstract might be located, it requires additional time to locate the full text article. Moreover, a majority of articles requested cost money to retrieve. A Medline search should be reserved for rare medical problems or research to write a paper or create, for example, a clinical practice guideline (CPG). Medline will be discussed in more detail in the chapter on search engines.

In 1994 Shaughnessy stated that the usefulness of medical information is equal to the relevance times validity divided by the amount of work to access it.[19] A recent study in the journal *Pediatrics* comparing retrieval of information from online versus paper resources showed it took eight minutes for an answer via an online resource as compared to twenty minutes using traditional paper based resources.[20] There is little doubt about the tremendous potential of online resources, in terms of quality and retrieval speed.

Harrison's Online Textbook of Medicine and The Scientific American (now known as ACP Medicine) were among the first online resources. They are updated more frequently than traditional textbooks and are accessible from anywhere. They generally cost about the same as print textbooks and are available online or on CD-ROMs. Their main drawback is that they tend to cover only the basics about any subject and therefore lack depth. Also, in spite of the fact that they have a search engine, like a standard textbook a reader may have to review multiple book chapters to find the answer.

More comprehensive but unfocused resources followed online textbooks. Examples are MDConsult, Medscape, StatRef, InfoRetriever and OVID. Most of these excellent products offer multiple resources such as books and journal articles. The chief limitation of these otherwise excellent resources is that fact that searches will likely lead to multiple results in books and journals. You might have to read twenty book or journal pages to finally find the answer. This is not optimal if you are seeking an answer while the patient is still in the exam room.

Ideal medical resources are those that are:
- Evidence based with references and level of evidence (explained in the chapter on evidence based medicine)
- Updated frequently
- Simple to access with a single sign-on
- Available at the point of care
- Capable of being embedded into an electronic health record
- Likely to produce an answer with only a few clicks
- Adequate for specialists <u>and</u> primary care physicians
- Written and organized with the end user in mind

According to Richard Smith the "best information sources provide relevant, valid material that can be accessed quickly and with minimal effort".[21] Such a resource would permit answering questions while the patient is still in the office, which is optimal. Several excellent and focused resources now exist. Examples of more focused resources include UpToDate, eMedicine, DynaMed, ACP-PIER and FirstConsult. How they differ is largely related to the number of topics covered as we will point out with each program. More focused programs such as UpToDate have been very well received.

- A 2004 study showed that 85% of medical students easily transitioned from traditional resources to primarily online medical resources (UpToDate and MDConsult) [22-23]
- In a report published in 2005, Internal Medicine residents were able to find answers 89% of the time and the information changed patient management 78% of the time. The most common resources accessed were UpToDate and Medline [24]

Several medical resource vendors are in the process of making the leap towards having the resource embedded into electronic health records. Examples would include iConsult, Dynamed, UpToDate and ACP-PIER to mention a few. The flow diagram below plots the evolution from the traditional to the online medical library.

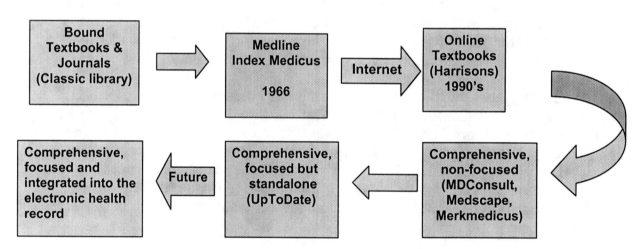

Figure 5.1. The evolution from traditional to online medical libraries

Free Commercial Medical Web Sites

Multiple excellent web sites are available that are either free or fee-based. Most of the sites discussed in this section have multiple features that continue to improve. Medical education traditionally has been based on reading journals or textbooks but can also involve the presentation of interesting and unique cases. A thorough discussion of this alternative approach appeared in the February 2007 Mayo Clinic Proceedings. [25]

Medscape (www.medscape.com)
- An all purpose medical web site

- Covers 20+ medical specialties as well as sections for nurses, medical students and pharmacists
- Provides updates, continuing medical education (CME), conference schedules, Medline, drug searches and multiple specialty articles
- Weekly newsletters and updates (MedPulse) and Best Evidence; both are features unique to Medscape
- Drug and Device Digest providing the latest in alerts and approvals; helpful for patient safety concerns
- A free personal web site option
- Dermatology atlas
- Clinical practice guidelines and Cochrane connection
- Con--some commercial influence [26]

MerckMedicus (www.merckmedicus.com)
- Multipurpose site
- 65+ specialty textbooks
- Customizable for 19 specialties
- 200+ full text journals
- Clinical podcasts
- Includes a limited version of MDConsult and OVID
- DxPlain differential diagnosis engine from Harvard
- TheraDoc antibiotic assistant for PDA
- Medical news including national meeting reports
- Mosby's drug database
- Patient handouts
- Professional development using CME, board reviews, medical meetings, medical school links and Braunwald's Atlas of Internal Medicine (1500 slides you can copy). Also, a slide image bank of other slides that can be copied
- PDA portal includes news, the Merck Manual, Pocket Guide to Diagnostic Tests, Journal abstracts and the ability to do a Medline search (if you are not wireless it will be done the next time you synchronize with your PC)
- Unique 3-D Atlas of the human body [27]

MedlinePlus (www.medlineplus.gov)
- Premier online patient education site
- Important to have in exam room
- Service of the National Library of Medicine and the National Institute of Health (NIH)
- Covers over 700 topics in English and Spanish
- Drug library
- Medical dictionary, encyclopedia and news
- 165 interactive video tutorials and surgical procedure videos

- Links to all major patient education sites: Mayo Clinic, National Institute of Health (NIH), etc
- Links to Clinical Trials.gov to search research centers for specific diseases [28]

Other Excellent Free Patient Education Sites
- www.familydoctor.org/
- www.mayoclinic.com/
- www.webmd.com/
- http://kidshealth.org/
- www.healthfinder.gov/

HighWire Press (http://highwire.stanford.edu)
- Free site created by Stanford University that hosts 919 journals with over one million articles
- One more alternative to OVID or trip to library [29]

E-medicine (www.emedicine.com)
- 6,000 articles by 10,000 authors. Covers more than primary care
- Monographs are free but other services are fee based
- Continually updated and peer reviewed. There are references but no footnotes in the text body. No level of evidence given
- 25,000 multimedia files
- Fee based CME (40,000 hours)
- Journals, images, drugs, articles, differential diagnoses and tools [30]

Amedeo (www.amedeo.com)
- This service will search major medical journals for
- a topic you select and then e-mail the results to you every week
- Valuable if you are a subject expert and don't have the time to do a frequent journal search on your own [31]
- Related to www.freebooks4doctors.com and www.freejournals4doctors.com
- Similar tracking of articles also available through Google Alerts and NCBI (Pubmed)

Subscription (fee based) Resources

Online Epocrates (www.epocrates.com)
- Online Epocrates was an obvious next step after the successful PDA software program (see chapter on mobile technology)
- The site organization is intuitive for clinicians
- You have an option to purchase your local formulary information as well as an extensive drug library to augment Epocrates

- In December 2005 a free online program was made available. In December 2006 pill pictures and patient education handouts were added to the free program
- Epocrates Linx is an online version that can integrate with an EHR
- Program covers 3300 drugs and 400 alternative medications
- The features online Epocrates has over the PDA version: Ability to print or e-mail results, Medline search capability, pill pictures, MedCalc 3000 calculations and patient education sheets in English and Spanish [32]

MicroMedex (www.micromedex.com)
- Micromedex has multiple databases. Toxicology = Poisondex, Disease = DiseaseDex, Lab = Lab advisor, Drug interactions = DrugDex
- Unlike Epocrates it has:
  - Both renal and liver failure dosing
  - Drug-food interactions
  - Off label uses
  - Comparative efficacy
  - Toxicology
  - References [33]

OVID (http://gateway.ovid.com)
- 38 textbooks covering most specialties
- Several hundred full text medical journals
- The ability to search multiple evidence based databases like Cochrane at the same time
- Medline search capability [34]

UpToDate (www.uptodate.com)
- Available online or on CD-ROMs or downloadable to PDA
- Updated each quarter as a rule
- Heavily weighted towards Internal Medicine
- No commercial backing
- Receive CME as you research a subject
- 3,000 authors review 290 journals
- Extensive peer review process
- **70,000 pages of text** and **7,000 topics**
- 19,000 graphics
- Links to 160,000 Medline abstracts
- Complete drug database to include drug-drug interactions
- Patient information topics
- Available for download to Pocket PC 2002 and 2003 but requires 2 GB of memory to download all features
- Now adding levels of evidence to articles
- Pediatrics, Neurology and Allergy/Immunology still in development
- Logically organized for fast answers

- Integrated into GE Centricity EHR [35]

MDConsult (www.mdconsult.com)
- 63+ textbooks
- Over 75 full text journals
- Comprehensive drug database
- 1000 clinical practice guidelines
- 2500 Patient education handouts
- Online CME
- Medline search capability
- MDC Mobile is the PDA portal
- Excellent Search engine for entire site [36]

StatRef (www.statref.com)
- Searches over 100 textbooks and Medline (depending on contract purchased)
- Also accesses ACP PIER (to be discussed)
- PDA portal will allow downloads of ACP PIER content only [37]

InfoRetriever/Infopoems (www.infopoems.com) A program that was created by physicians for physicians. POEMS are patient oriented evidence that matters. Specifically, this means the authors look for articles that are highly pertinent to patient care and patient outcomes
- Consists of two products: DailyPOEMS and InfoRetriever
- InfoRetriever searches multiple resources
- DailyPOEMs are e-mailed to you M-F and are distilled from 100+ journals with only 1/40 accepted
- Site has 2000 POEMS
- Cochrane abstracts (2,193)
- Practice Guidelines (751)
- Clinical decision rules (231)
- Number Needed to Treat (NNT) tool
- Photographic atlas
- Available for desktop or on Palm or Pocket PC PDAs
- Diagnosis calculators (1180)
- History and physical exam calculators (1282)
- 5 Minute Clinical Consultant
- ICD-9 and E&M lookup tool
- Drug of Choice tool
- Searching results in a summary of resources on that topic categorized into typical quick reference categories like diagnosis, treatment, prognosis, etc. 5 Minute Clinical Consult monographs are listed first[38]

ACP Medicine (www.acpmedicine.com)

- Publication of the American College of Physicians
- Previously known as *Scientific American Medicine*
- Covers most subspecialties plus Psychiatry, Women's Health, Dermatology and Interdisciplinary medicine
- Available in binder, CD's and Online
- Up to 120 hours CME available
- Binder version is 2500 pages
- Articles are dated and references are footnoted with PubMed links to the abstract
- Monthly updates (free) to be added to chapters [39]

ACP PIER (http://pier.acponline.com)

- About 430 topics that have a level of evidence and references
- What they cover they do well. Like an online textbook and updated frequently
- PDA version available
- Also includes drug database and procedures
- Provides the medical resource content for Allscript's EHR
- Modules continue to be added [40]

FirstConsult (www.firstconsult.com)

- Synthesizes evidence from journals and other sources into one database
- Organized into differential diagnoses and topic searches
- 475 topics at this point
- Differential diagnosis generator
- PDA Portal
- 30 day free trial
- 300 Patient education files in English and Spanish
- Procedure files and videos
- Handheld option, but for full library of topics it requires 10MB memory
- Negatives are no drug database and limited topics [41]

DynaMed (www.dynamicmedical.com)

- Disease and condition reference
- 2000 clinical topics
- Weekly e-mail of important articles; also available as podcast
- Constantly updated. Level of evidence included
- Linked to articles, handouts and guidelines online
- PDA version free with subscription
- All topics are organized in the same categories such as, general information, causes and risk factors, complications and associated conditions, history, physical, diagnosis, prognosis and treatment

- Bottom line recommendations are presented first, along with level of evidence. Links to articles will take you to the full text article if available and free online. Other links take you to PUBMED where some are linked through medical libraries to full text articles.[42]

Table 5.1 is a matrix that compares many of the features of the online resources just covered. The speed of retrieval is an approximate estimate of how much time it takes to find an answer to a common medical question.

Source	Medline	CME	Books	Journals	Drugs	News Letters Updates (e-mail)	Expert Opinion	Patient Info	Speed (1-4)
Medscape	X	X	X		X	X	X	X	3
MerckMedicus	X	X	X	X	X		X	X	3
OVID	X		X	X					2
UpToDate		X			X		X	X	4
MDConsult	X	X	X	X	X		X	X	2
FirstConsult							X	X	4
StatRef	X		X						2
ACP Medicine		X				Updates To site only	X		3
eMedicine	X	X	X	X	X		X	X	3
DynaMed					X	X	X	X	4

Table 5.1. Online resource comparison matrix. Speed (Slow =1, Fast = 4)

Conclusion

Online resources are becoming the medical library of choice for healthcare workers due to depth of content and speed of retrieval. Furthermore, subject matter can be updated more rapidly compared to standard textbooks. Many excellent resources are free and the subscription resources are competitive with traditional textbooks. Resources vary from a low of about 400 topics to a high of 8000 topics. Prices tend to correlate with the size of the content offered. There are many free resources that should be considered by all clinicians such as Epocrates Online, MedlinePlus and Medscape. The authors want to stress that very extensive resources such as UpToDate, E-medicine and DynaMed offer the greatest possibility of finding an answer in a few clicks. Other resources may point you to multiple book chapters and journal articles where you must sift through the data to find the answer. Clinicians are strongly encouraged to "test drive" these resources, adopt the ones that make the most sense and add them as desktop icons in each exam room.

References

1. Medline Fact Sheet http://www.nlm.nih.gov/archive/20041201/pubs/factsheets/medline.html (Accessed May 5 2006)
2. Physician Desk Reference http://www.pdrbookstore.com/Merchant2/merchant.mv?Screen=PROD&Store_Code=001&Product_Code=160002 (Accessed May 7 2006)
3. Richardson WS, Wilson MC Textbook descriptions of disease—where's the beef? ACP Journal Club July/August 2002: A-11-12
4. Shaneyfelt T. Building bridges to quality. JAMA 2001;286:2600-01
5. Contopoulos-Ioannidis DG, Ntzani E, Ioannidis JP. Translation of highly promising basic science research into clinical applications. Am J Med. 2003 Apr 15;114(6):477-84.
6. Antman EM et al A comparison of results of meta analyses of randomized control trials and recommendations of clinical experts JAMA 1992;268:240-248
7. Ioannidis JPA Contradicted and Initially Stronger Effects in Highly Cited Clinical Research JAMA 2005;294:219-228
8. Ramsey PG et al Changes over time in the knowledge base of practicing internists JAMA 1991;266:1103-1107
9. Covell, DG, Umann GC, Manning PR. Information needs in office practice: are they being met? Ann Int Med 1985;103:596-599
10. Ely J, Osheroff J, Ebel M et al Analysis of questions asked by family doctors regarding patient care BMJ 1999;319:358-361
11. Gorman PN, Helfand M How physicians Choose Which Clinical Questions to Pursue and Which to Leave Unanswered. Med Decision Making 1995;15:113-119
12. Ely JW et al Analysis of questions asked by Family doctors regarding patient care BMJ 1999;319:358-361
13. Lee T Quiet in the Library NEJM 2005;352: 1068-70
14. Lindberg DAB Humphreys BL 2015-The Future of Medical Libraries NEJM 2005;352:1067-1070
15. Technology usage in physician practice management AMA survey Dec 2001
16. Miccioli G Researching Medical literature on the Internet---2005 Update www.llrx.com/features/medical2005.htm (Accessed December 10 2005)
17. Gorman PN. Can primary care physicians' questions be answered using the medical journal literature? Bull Med Libr Assoc 1994;82:140-146
18. Chambliss ML Answering clinical questions J Fam Pract 1996;43:140-144.
19. Shaughessy A, Slawson D, Bennett J Becoming an Information Master: A guidebook to the Medical Information Jungle J Fam Pract 1994;39:489-499
20. D'Alessandro DM, Kreiter CD and Petersen MW An Evaluation of Information Seeking Behaviors of General Pediatricians Pediatrics 2004;113:64-69
21. Smith R What Clinical Information do Doctors Need? BMJ 1996;313:1062-1068
22. Peterson MW et al Medical student's use of information resources: is the digital age dawning? Acad Med 2004;79:89-95
23. Schilling LM et al Residents' patient specific clinical questions: opportunities for evidence based learning Acad Med 2005;80:51-56

24. Massachusetts General Hospital Study. Available at www.uptodate.com (Accessed January 1 2006)
25. Pappas G, Falagas ME Free Internal Medicine Case-based education through the World Wide Web: How, Where and With What? Mayo Clin Proc 2007;82(2):203-207
26. Medscape www.medscape.com (Accessed January 28 2007)
27. Merck Medicus www.merckmedicus.com (Accessed January 28 2007)
28. MedlinePlus www.medlineplus.com (Accessed January 20 2007)
29. High Wire Press http://highwire.stanford.edu (Accessed January 15 2006)
30. E-Medicine www.emedicine.com (Accessed January 14 2007)
31. Amedeo www.amedeo.com (Accessed January 27 2006)
32. Epocrates www.epocrates.com (Accessed January 1 2007)
33. Micromedex www.micromedex.com (Accessed February 1 2006)
34. OVID http://gateway.ovid.com (Accessed February 1 2006)
35. UpToDate www.uptodate.com (Accessed February 3 2007)
36. MDConsult www.mdconsult.com (Accessed February 2 2006)
37. StatRef www.statref.com (Accessed January 15 2006)
38. Inforetriever www.infopoems.com (Accessed January 14 2007)
39. ACP Medicine www.acpmedicine.com (Accessed January 24 2006)
40. ACP Pier http://pier.acponline.org (Accessed January 20 2006)
41. First Consult www.firstconsult.com (Accessed February 4 2006)
42. DynaMed www.dynamicmedical.com (Accessed February 15 2007)

 Chapter 6: Search Engines

Learning Objectives

After reading this chapter the reader should be able to:
- State the significance of rapid high quality medical searches
- Define the role of Google in healthcare and its many search features
- Describe the meta-search engines and the features that are distinct from Google
- Describe the role of PubMed and Medline searches
- Identify the variety of search filters essential to an excellent PubMed search

Introduction

> "Getting information off the Internet is like taking a drink from a fire hydrant"
>
> Mitchell Kapor

The most rapid and comprehensive way to access information today from anywhere in the world is a search of the World Wide Web via the Internet. If we assume that the Internet is the new global library with more than 3 billion web sites, then it should come as no surprise that search engines are the gateway. Popular search engines such as Google provide successful searches for medical and non-medical issues. Although PubMed is the search engine of choice for formal searches of the medical literature, most inquires are informal so searches need to yield primarily rapid and relevant results. Given the prevalence of web surfing for answers, multiple articles have been written about "search wars".[1-2] It is unclear to what extent the use of search engines has changed human behavior and medical knowledge in the past decade. Previously, questions such as what is the difference between HDL and LDL cholesterol meant a trip to the library, the purchase of a book or a visit with a doctor. Now anyone can execute a search and have a reasonable likelihood that the search will be successful.

Just as important as selecting a search engine you are comfortable with is learning to use all of the advanced search options. It is imperative to use filters to refine a search or you will become frustrated by the avalanche of information returned. In this chapter we will begin with a discussion of Google, followed by other less well known search engines and finally a primer on PubMed searches.

Google

Google is by far the most widely used search engine in the world.[3] Its name is derived from the word googol which is the mathematical term for the number 1 followed by 100 zeroes. [4] Google's success is based largely on its intuitiveness, retrieval speed and productive results. Google is listed as one of the ten forces to flatten the world in Thomas Friedman's book "The World is Flat".[5] Google has proven to be a fascinating company with a myriad of innovations on a regular basis.

Google was developed by Larry Page and Sergey Brin in 1996 when they were graduate students at Stanford. They created the "backrub" strategy which meant that a search would prioritize the results by ranking the page that is linked the most first (page ranking).[6] Some could argue that it used a popularity contest as a strategy. A shortcoming of this approach would be that new web sites might take time to be linked. As the worlds largest and fastest search engine it performs one billion searches daily by utilizing thousands of servers (server farm) running the Linux operating system.[7] Google can be criticized for being a shotgun and not a rifle in terms of returning too many results but this has not diminished its popularity. Because Google yields so many results in an average search it is very important to learn about how to narrow or filter a search.

Google can be an acceptable medical search engine for common as well as rare conditions not likely to be found in journals or textbooks. Google provides a very global review, returning articles from the lay press, medical journals, magazines, etc. It is important to point out that Google will cite Medline abstracts and occasionally full text articles, so for an informal search it isn't unreasonable to start with Google to see if you find an answer in the first few citations listed. It is likely you will find an acceptable answer in less time than it takes to use PubMed, particularly if you use an advanced search strategy. Adding additional descriptors to the search is essential. As an example, if you search with the terms _type 2 diabetes foot checks frequency_ you will likely retrieve clinical articles that describe how often foot checks should be performed in diabetics. The most important term should be listed first. In a 2006 article, Dr. Robert Steinbrook notes that Google (56%) was the most common search engine used to refer someone to find a medical article at High Wire Press; compared to PubMed (8.7%).[8] Several recent articles in the medical literature have confirmed that Google has become a common medical search engine, even at academic centers.[9-11] We will list several of the Google options essential for a successful search:
- Begin by setting preferences

- o Language you prefer e.g. English
- o Number of search results per page e.g. 10 or 20
- o Whether you want your search to launch in new window (recommended because should you exit the current page, you won't exit the Internet)
- Select the *Advanced search* option on the main page:
  - o Under occurrences you can search for a term in the title only or in the body or both
  - o Select search by domain such as .org or .gov
  - o Select search by format: Word, Excel, PDF, PowerPoint, etc
  - o It narrows the search if you put quotation marks around the words e.g. "University of West Florida" so you don't retrieve every citation with the words Florida, West or University
- Take a look at *Advanced operators* to refine the search
  - o One that is particularly helpful is *define*: type a word in after the colon and it serves as a dictionary
- **Essentials of a Google Search** (www.google.com/help/basics.html) Provides helpful tips to improve the search process:
  - o Searches are not case sensitive
  - o The word "and" is not necessary because Google, unlike PubMed, does not use Boolean operators, with the exception of using the word OR
  - o Most popular web sites are listed first
  - o For a search of a common subject, select "I'm feeling lucky" on the main Google search page and you will be taken to the most popular web sites on the subject [12]

The following are Google features that, in our judgment are relevant to Medical Informatics and available as of March 2007:

- Google includes an *Image* search of over 880 million images (some copyrighted). Advanced search filters are available
- Google changes so rapidly it is a good idea to look at *Google labs* often. It provides new web page alerts and news alerts, toolbar shortcuts, a glossary and discussion groups and the ability to create a web page. Alerts can track any topic and e-mail you any new information as it becomes available.
- *Google Talk* is voice over Internet protocol (IP). You can talk to another person via your computer and the Internet
- *Gmail* is free web mail. Unlike other web mail services it offers over 2 gigabytes of memory. Google maintains that you don't need to organize your e-mail into folders because of its excellent search engine
- *Google Docs & Spreadsheets* provides an alternative to Microsoft Word and Excel. You can collaborate with others and publish your work to a web site

- *Google Groups* creates a collaborative web site where you can post discussions and web content. Members can be by invitation only
- *Google Directory* organizes the Web into categories so the search may be more focused. If you search, for example, under a Health directory > Medicine > Informatics > Telemedicine, the search will yield page ranked web sites so you do not see citations from the lay press, PubMed, etc. The web site choices are actually selected by experts so this is why the number of sites returned is far smaller than a true web search
- *Google Health* is a new beta feature that allows searching for issues under categories such as overview, treatment, diagnosis, clinical guidelines, symptoms, complications and many more [13]
- *Google Custom Search Engines* under "co-op" you will find the ability to customize searches by limiting them to certain web sites. You can also integrate this search engine into your personal web site
- *Google Code* searches for open source codes and APIs
- *Google Page Creator* creates simple web pages with easy to use tools so anyone can create their own web site
- The *Scholar* feature searches authors and their publications with *advanced search* and *help options* [14-15]
- The *desktop search* program rapidly searches your personal computer files. Tends to be much faster than the Microsoft search function located in the start menu
- *Book Search (beta)* is a program that potentially will search 50 million textbooks. If a book is currently copyrighted you will only see a sample page. Further description: *"When we find a book whose content contains a match for your search terms, we'll link to it in your search results. Click a book title and you'll see the Snippet View which, like a card catalog, shows information about the book plus a few snippets - a few sentences of your search term in context. You may also see the Sample Pages View if the publisher or author has given us permission or the Full Book View if the book is out of copyright. In all cases, you'll also see 'Buy this Book' links that lead directly to online bookstores where you can buy the book."* [16]

Other Search Engines

Meta-search engines search more than one database or utilize more than one search engine.[17-20] It remains to be seen if this is necessarily an advantage or not. We could find no publications or reviews regarding their use in medical searches.

A9
Other search engines are available that offer features that set them apart from Google. One such search engine is A9.

- Located on the main page of Amazon.com
- Adds structure to a search. You can create categories (in columns) on your main results page: web, movies, books, images, blogs, yellow pages, people, references and Wikipedia
- Under "other choices" is PubMed and more than 200 web sites you can add for specific searches
- Place cursor over "Site info" and you will see how site is ranked and how many links it has
- Note the small sponsored section
- You can create an A9 toolbar like Google [21]

Vivisimo (www.vivisimo.com)
- Search technique developed by Carnegie Mellon researchers
- Free commercial product available to search an enterprise's Intranet and Internet
- Advanced search option appears after you have performed the basic search and only includes the option to select different databases
- "Clusters" the search into convenient folders on the left
- Adopted by the National Library of Medicine and MedlinePlus to use on their web sites
- Vivisimo now offers a *Velocity* option that is more specific for the life sciences and bioinformatics [22]

Dogpile (www.dogpile.com)
- Searches Google, MSN, Ask Jeeves and Yahoo
- Advanced search uses the filters of qualified words or phrases, language, dates and domain; similar to Google
- Searches sponsored and non-sponsored web sites
- You can separately search the web, images, audio, video, news, white and yellow pages
- Note: a 2006 search for avian influenza returned 98 high quality citations, whereas a Google search returned 20 million! [23]

Omni Medical Search (www.omnimedicalsearch.com)
- Searches 15 medical sites such as the Centers for Disease Control (CDC) and WebMD
- Can search the web, news, images and Web 2 (academic and government domains)
- MedPro search is intended for serious medical inquiries, yet yielded very limited results and is associated with commercial influence. A 2006 search, as an example, for avian influenza yielded only 49 results [24]

PubMed Search Engine

Anyone seeking information from the medical literature should know how to use PubMed. This is particularly true in an academic or research environment. If the intention of the search is to look for an answer to a relatively simple question, then PubMed is probably not the search engine of choice. Moreover, without proper training a search can be challenging and frustrating. While some would argue that all physicians should learn to use PubMed to retrieve evidence based medical answers, others would argue the process is too labor intensive. This section was written to emphasize important features and shortcuts to make a search easier and more successful. In addition, excellent tutorials exist on the PubMed site to teach you the basics of a good search. Also, several helpful review articles have been written that address PubMed tools and features. [25-27]

PubMed is a web-based retrieval system developed by the National Center for Biotechnology Information (NCBI) at the National Library of Medicine (NLM). Since its beginning in 1996, use has steadily grown to about 70 million searches in 2005.[28] It is part of NCBI's vast retrieval system, known as *Entrez*. Its primary purpose is to search the database known as MEDLINE that includes citations from the world's medical literature.[29]

MEDLINE is the NLM's premier bibliographic database covering the fields of medicine, nursing, dentistry, veterinary medicine, the health care system, the preclinical sciences and some other areas of the life sciences. MEDLINE contains bibliographic citations and author abstracts from over 4,800 journals published in the United States and 70 foreign countries.[30] MEDLINE has over 13 million records from 1966 to the present. Below is an example of a MEDLINE citation listing the author, title, journal, published date and PubMed identification number (PMID) (Figure 6.1).

Zerboni L, Sommer M, Ware CF, Arvin AM.
Varicella-zoster virus infection of a human CD4-positive T-cell line.
Virology. 2000 May 10;270(2):278-85.
PMID 10792986 [PubMed - indexed for MEDLINE]

Figure 6.1. Medline citation (courtesy National Library of Medicine)

It is important to realize that PUBMED is only one out of twenty three databases available from the NLM and the National Institute of Health. The others are primarily biological databases that are part of bioinformatics and genomics. A link to free online biomedical textbooks is also included.

Medical Subject Heading (MESH). Perhaps the first concept to understand prior to performing a search is MESH. Journal abstracts have to be categorized by NLM indexers in order for them to be searched. Articles are saved under one or more subject headings using a structured vocabulary. As you can imagine, terms such as low back pain could be labeled lumbar pain, osteoarthritis of the lumbar spine, etc. Figure 6.2 shows how sinusitis is organized using MESH terms. In spite of the hard work by the indexers, you will occasionally see citations that seem to have little to do with what you are requesting. You may want to check your search terms first under MESH to see if the term is accepted by PUBMED. You can access it in the drop down menu in the search window or by choosing the MESH Database in the menu on the left.

All MeSH Categories
    Diseases Category
        Respiratory Tract Diseases
           Nose Diseases
             Paranasal Sinus Diseases
              Sinusitis
                Ethmoid Sinusitis
                Frontal Sinusitis
                Maxillary Sinusitis
                Sphenoid Sinusitis

Figure 6.2. MESH categories

In figure 6.3 we searched to see if sinusitis was an acceptable term and looked for alternatives. Different types of sinusitis are listed and similar terms are listed under "suggestions".

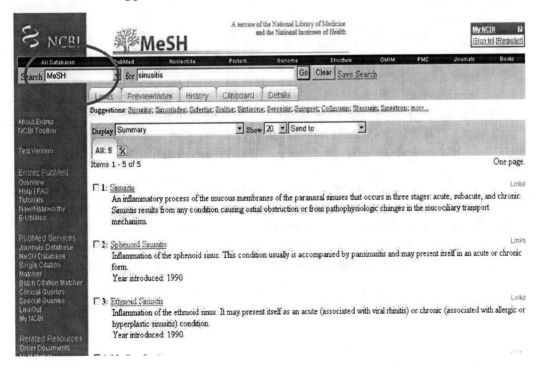

Figure 6.3. MESH search (courtesy National Library of Medicine)

If you scroll down the drop down search menu and change the search from MESH to PubMed and search for sinusitis you discover over 13,000 citations. Like a Google search, learning to use filters will result in more successful retrievals of information.

PubMed Limits. This section of PubMed underwent significant changes in March 2006.
- You can now search by author or journal
- You can also search for full text and free full text articles and abstracts. Keep in mind that most articles before 1975 did not contain abstracts
- You can continue to search by date, age of subjects, gender, humans or animals, language, publication types, topics and field tags
- Searchable main publication types:
  - Clinical Trial
  - Editorial
  - Letter
  - Meta-Analysis
  - Practice Guideline
  - Randomized Controlled Trial
  - Review
- Searchable topic subsets:
  - AIDS
  - Bioethics
  - Cancer
  - Complementary Medicine
  - History of Medicine
  - Space Life Sciences
  - Systematic Reviews
  - Toxicology
- Field tags. You can stipulate whether you want the search term in the title or body of the article. Multiple other choices are listed as well

We are now going to search with the following limits for sinusitis: Field = title, language = English, Age = 19-44, humans, Medline, added to PubMed in past 5 years and links to free full text. We elected to search for a practice guideline that would give us recommendations about treatment. We could have also selected clinical trial, random controlled trial or review or checked the box for all four.

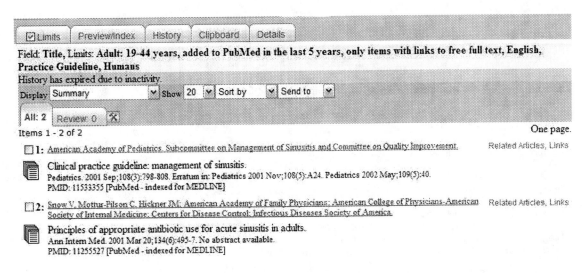

Figure 6.4. Search for sinusitis with multiple limits (courtesy National Library of Medicine)

Our search with limits has greatly reduced the number of returned citations and improved the quality (Fig. 6.4). Requesting free full text articles also reduces the search considerably. When we changed the search to any abstract, instead of free full text articles, 60 citations were returned.

- Note that the most current articles are listed first
- Also notice that in spite of asking for articles only dealing with adults, a pediatric guideline returned. Perhaps this is because the guideline deals with patients ages 1-21.
- To the left of the article you will see icons that have the following significance (Fig. 6.5):

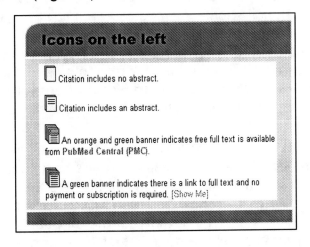

Figure 6.5. Icons (courtesy National Library of Medicine)

- Most articles today are associated with an abstract that summarizes the article (Fig.6.6)
- You must go to the full text article for more detail

☐ **1:** Am J Rhinol. 2006 Nov-Dec;20(6):658-61.

Making the call: the diagnosis of acute community-acquired bacterial sinusitis.

Le Annie V, Simon RA.

From the Division of Allergy and Immunology, The Scripps Clinic and Research Institute, La Jolla, California 92037, USA. annievole@yahoo.com

BACKGROUND: Although one of the most common illnesses encountered in the primary care setting, acute community-acquired bacterial sinusitis (ACABS) can be a challenge to diagnose. METHODS: Existing diagnostic modalities ranging from clinical history to imaging studies used to diagnose ACABS are discussed. RESULTS: Numerous methods exist but they do not distinguish well between viral and bacterial illness. CONCLUSION: Diagnosis of ACABS should primarily be made based on the clinical history. Other modalities provide useful information in select cases.

PMID: 17181113 [PubMed - in process]

Figure 6.6. Example of an abstract (courtesy National Library of Medicine)

- Pubmed uses Boolean operators such as AND, OR and NOT and they must be capitalized. Instead of the search strategy used above we could have used the same limits but in the main search window used "sinusitis AND treatment" (no quotation marks necessary)
- Other tab options

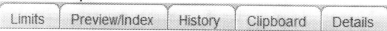

  - Preview/Index--preview the number of search results before displaying the citations. View your search strategy as you continue to refine your search
  - History--holds all of your searches for 8 hours. You can combine several searches by combining for example #2 and #3 searches
  - Clipboard--allows you store up to 500 citations for up to 8 hours
  - Details--summarizes how you performed your search
- More options

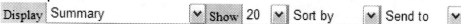

  - Display window--it automatically defaults to summary but you can also select abstract, Medline, XML and others
  - Sho--you can elect to show between 5-500 citations per page
  - Sort by-- you can sort by author, journal or published date
  - Send to--your choices are e-mail, text, file, printer, clipboard, RSS feed or order article from a commercial site (Loansome Doc)
- Single Citation Matcher—located under the left main menu. There are times when you are trying to locate an article and you have an approximate idea as to who the author is or the year or the journal. You can search by author, journal, date, volume, issue, page or title words

- Clinical Queries—also located under the left main menu. Provides another way to search for articles with some built in filtering.
  - Clinical study category—permits searching in a broad/sensitive manner or narrow/specific manner. Note that you can search by etiology (cause), diagnosis, therapy, prognosis or clinical prediction guides
  - Systematic reviews—this type of review critically appraises multiple random controlled trials to give conclusions more strength (covered in more detail in the chapter on evidence based medicine
- My NCBI—also located on left main menu. Provides storage area for articles you are retrieving
  - Save searches (otherwise gone in 8 hrs)
  - Set up e-mail alerts so you are notified when new articles are published on your topic of interest
  - Display links to Web resources (LinkOut)
  - Choose filters that group search results
  - You must first register for this free service
- Related articles and Links. To the right of each article you will see a hyperlink to similar or related articles. Select Links and it can link you to:
  - PubMed books
  - Links—if the article is free and full text, PubMed Central may have it
  - Link Out—can take you to external resources such as OVID or MDConsult if you subscribe and set this up. May also take you to other sites such as MedlinePlus for patient education material
- PubMed PICO Option—it is important to construct a well thought out question prior to initiating a search. With this tool it divides the question into sections P.I.C.O. The URL or web address could be a desktop icon shortcut for fast searches.[31]
  - (P) atient or problem—how do you describe the patient group you are interested in? Elderly? Gender?
  - (I) ntervention, prognostic factor or exposure-- Drug? Lab test? Tobacco?
  - (C) omparison—with another drug or placebo?
  - (O) utcome—what are you trying to measure? Mortality? Reduced heart attacks?

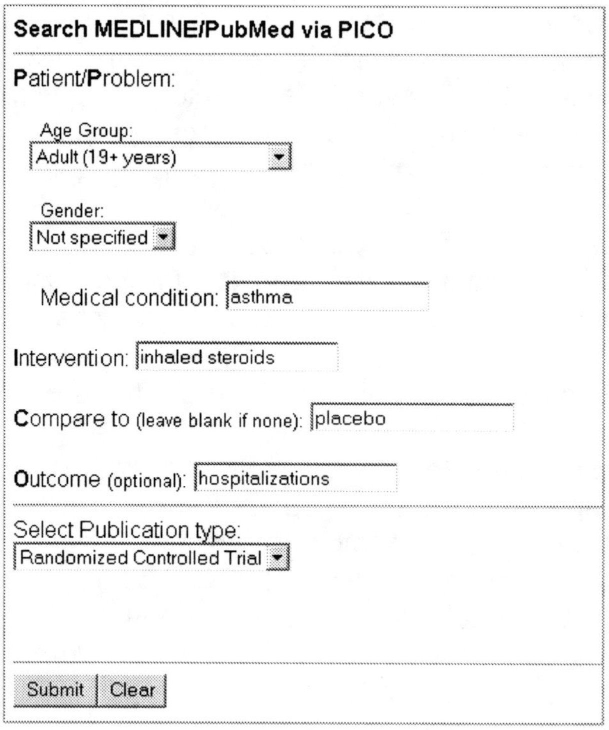

```
Search MEDLINE/PubMed via PICO

Patient/Problem:

    Age Group:
    [Adult (19+ years)        ▼]

    Gender:
    [Not specified ▼]

    Medical condition: [asthma          ]

Intervention: [inhaled steroids      ]

Compare to (leave blank if none): [placebo        ]

Outcome (optional): [hospitalizations     ]

Select Publication type:
[Randomized Controlled Trial ▼]


[Submit] [Clear]
```

Figure 6.7. PICO search (courtesy National Library of Medicine)

In the example we searched based on the question "do inhaled steroids in adults with asthma decrease hospitalizations", looking only at random controlled trials. The search returned several very relevant articles (Fig. 6.7).

- PubMed Central—hosts multiple free full text articles. Unfortunately, many are located in minor journals of recent vintage. They are also more weighted towards a bioinformatics search.

Conclusion

At this time, Google is the premier search engine for non-medical and perhaps medical searches. With proper filtering and experience, Google can be used with significant success. Better studies are needed to compare Google with other search engines and PubMed. Familiarity with PubMed and its new features is important for healthcare workers who need to conduct formal searches of the medical literature. With knowledge and experience a PubMed search can result in relevant results in a timely fashion.

References

1. Ferrara F The Search Wars MD Net Sept 2004:28-30
2. Al-Ubaydli Using search engines to find online medical information   PLOS Medicine 2005 http://medicine.plosjournals.org (December 13 2005)
3. Standard & Poors Insight/Express 2004 survey. www.emarketer.com (Accessed March 14 2006)
4. Google.pedia. The ultimate Google resource. Michael Miller. Que publishing. 2007
5. Friedman, Thomas. The World is Flat. Farrar, Straus and Giroux. New York. 2006
6. The Anatomy of a large scale hypertextual web search engine http://www.db.stanford.edu/~backrub/google.htm (Accessed March 14 2006)
7. Wickipedia: Google http://en.wikipedia.org/wiki/Google (Accessed March 14 2006)
8. Steinbrook R Searching for the Right Search—Reaching the Medical Literature NEJM 2006;354:4-7
9. Tang H, Ng JHK. Googling for a diagnosis—use of Google as a diagnostic aid: Internet study BMJ Nov 11 2006 www.bmj.com (Accessed November 15 2006)
10. Correspondence. And a Diagnostic Test was Performed. NEJM 2005;353:2089-2090
11. Turner MJ Accidental epipen injection into a digit—the value of a Google search Ann R Coll Surg Engl.2004;86:218-9
12. Essentials of a Google Search. http://www.google.com/help/basics.html (Accessed November 1 2006)
13. Google Health www.google.com/health (Accessed March 1 2007)
14. Butler D Science searches shift up a gear as Google Starts Scholar engine www.nature.com 2004;432:421 (Accessed March 14 2006)
15. Google Scholar Beta Peters Digital Reference Shelf Thomson-Gale www.galegroup.com December 2004 (Accessed February 4 2005)
16. Google Book Search http://books.google.com/ (Accessed March 26 2006)
17. Patterson B Searching beyond Google and Yahoo: nine online search engines compared May 9 2005 http://reviews.cnet.com (Accessed May 25 2005)
18. Search Wars http://news.bbc.co.uk/1/hi/magazine/4003193.stm (Accessed September 9 2006)
19. Brandon J. Outside the box: search the web with power PC Today Dec 2004: 39-41
20. Nelson R. Smart Web Surfing. Physicians Practice Sept 2004: 73-75
21. A9 Search Engine www.amazon.com (Accessed March 14 2006)
22. Vivisimo www.vivisimo.com (Accessed March 14 2006
23. Dogpile www.dogpile.com (Accessed March 14 2006
24. Omni Medical Search www.omnimedicalsearch.com (Accessed March 14 2006)
25. Ebbert JO, Dupras DM, Erwin PJ Searching the Medical Literature Using PubMed: A tutorial Mayo Clin Proc. 2003; 78:87-91
26. Sood A, Erwin PJ, Ebbert JO Using Advanced Search Tools on PubMed for Citation Retrieval Mayo Clin Proc. 2004;79:1295-1300
27. Haynes RB, Wilczynski N Finding the gold in MEDLINE: Clinical Queries ACP Journal Club Jan/Feb 2005;142: A8-A9
28. Steinbrook R Searching for the Right Search----Reaching the Medical Literature NEJM 2006;354:4-7
29. Entrez PubMed www.pubmed.gov (Accessed March 14 2006)
30. Fact Sheet MEDLINE www.nlm.nih.gov/pubs/factsheets/medline.html (Accessed September 1 2004)
31. NLM PICO http://askmedline.nlm.nih.gov/ask/pico.php (Accessed March 20 2006)

Chapter 7: Mobile Technology

Learning Objectives

After reading this chapter the reader should be able to:
* Describe the history behind mobile technology
* List the essential features of a personal digital assistant
* Identify the limitations of hand held technology
* Compare and contrast the medical software programs most helpful for clinicians
* Describe the evolution from personal digital assistants to PDA phones

Introduction

Had this section been written just one year ago it would have been given the title "Personal Digital Assistants and Medicine". With the advent of PDA phones, BlackBerrries and "smart phones", we feel that mobile technology is a more appropriate title. Mobile technology is a very logical transitional step from the personal computer. With improving speed, memory and shrinking size, consumers desire a more portable platform for their information and applications. Although mobile technology is not necessarily part of a Medical Informatics book or course, widespread use and popularity of this technology in medicine make it worth mentioning. Additionally, one of the authors (RH) has extensive experience with PDAs and medical education.

The history of personal digital assistants is quite recent. In the early 1990's the Apple Newton appeared and held great hope that we would have handheld technology that would appeal to the average user (Figure 7.1). This monochrome PDA weighed .9 lbs, measured 7.25 x 4.5 x .75 inches, had 150 K of SRAM, a processor speed of only 20 mHz, short battery life and cost $700.[1] It obviously did not succeed because it was too big, too slow, had insufficient memory, was too heavy and cost too much for the average consumer.

Figure 7.1. AppleNewton

The next handheld product to catch the public's attention was the Palm Pilot 1000, invented by Jeff Hawkins in 1994 and released in 1996.[2] It was smaller, less expensive and had 128 K of memory. Synchronizing with the personal computer was a one-step operation.

One could argue that the PDA did not become popular with the medical profession until the "killer application" Epocrates was released in 1999.[3] First, there was the excitement of knowing that drug facts could be retrieved much more rapidly with the PDA compared to the Physician Desk Reference (PDR) and secondly, the product was free. The PDA was also a platform to store all medical "pearls" rather than stuffing more notes into the pocket. Slowly, other companies got on the bandwagon to produce PDAs; like Hewlett-Packard (iPaq), Dell (Axium) and Sony (which has since abandoned the PDA market). Microsoft's Windows operating system (OS) has made major inroads into what initially was mainly Palm OS territory for the medical profession. Symbian and Linux are also operating systems that can be used in the PDA but are not prevalent in the United State's PDA market. BlackBerry by Research in Motion is extremely popular for e-mail and recently medical software programs, to include e-prescribing, are beginning to appear.[4] BlackBerry programs will be covered later in this chapter.

Figure 7.2 Palm Pilot

PDA Popularity

As of 2004, about 30,000,000 Palm PDAs had been sold worldwide. According to a 2002 article in the Journal of the American Medical Informatics Association, 66% of Family Practice residency programs use PDAs, with the highest user rate reported in the military at 80%.[5] Many medical schools fully support handheld technology as part of their educational programs. Figure 7.3 shows the adoption rate of mobile devices based on four surveys.[6] PDA use by clinicians has almost quadrupled in six years.

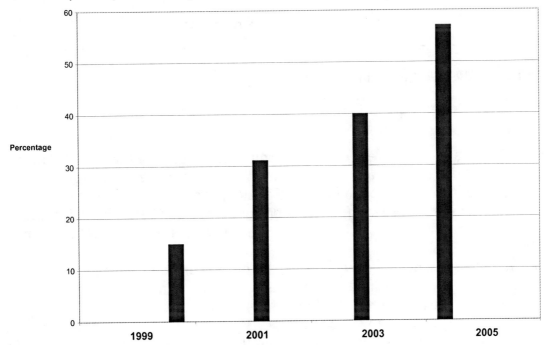

Figure 7.3 Mobile Device Adoption by Physicians

The following figure was adapted from a 2004 article in the journal Pediatrics, shows the most common uses of PDAs by Pediatricians (Figure 7.4).[7] As other studies have shown, a drug reference is listed as the most frequently used program. Personal scheduling is listed second because most people need to track their busy schedules and most of these program sync with Microsoft Outlook. Perhaps this category also included contact information, as that is also a major benefit of handheld technology. Note that most people did not use a PDA to: store patient information, e-prescribe, access the Internet or submit a bill. The PDA does not excel in those areas due to a small screen size, an inadequate keyboard and the need for more complicated and expensive software. It also requires more complex training that is difficult to achieve in a private practice setting. On the other hand, PDAs are great for simple medical calculations.

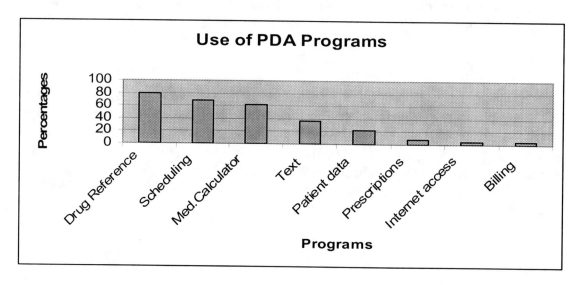

Figure 7.4. Use of PDA Programs

Basic PDA Functions

Most PDAs have four lower buttons for *calendar* (date book), *contacts* (address book), *tasks* (to do list) and *memos* (notes). These are the basic organizer functions one expects from a PDA. In addition, users will usually have a calculator, expense program, syncing option, e-mail, infrared beaming ability, notepad and an alarm clock application. Two useful functions related to these programs are:

1. When you establish a new event in the date book you have the option of setting an alarm as a reminder and it can be set up as a recurring event. If you have a staff meeting the fourth Wednesday of each month, your PDA will remind you with an alarm every month.

2. You can create an e-business card for all of your contact information. When you select and hold down the contact button for two seconds it will

automatically beam an electronic business card to your associate's PDA instead of handing him/her a business card.

All of these functions synchronize to your personal information manager (PIM) that could be Microsoft Outlook or the Palm desktop software. Below are screen shots of the calendar and contact features of the Windows operating system (Figure 7.5).

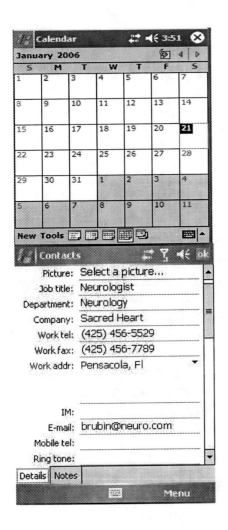

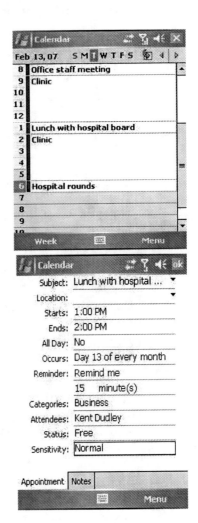

Figure 7.5. Windows OS calendar and contact features

Medical PDA Programs

Calculators

1. **Cardiac Clearance.** A free program that offers two nationally recognized algorithms for cardiac clearance before surgery. Simply tap on a few screens to answer standard questions and it tells you if the patient needs further testing prior to going to the operating room. [8]

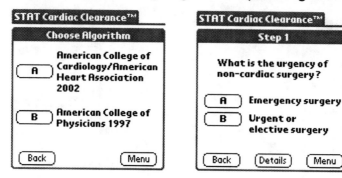

Figure 7.6. Cardiac Clearance

2. **Cholesterol.** Software is based on the national Adult Treatment Panel (ATP) III guidelines. This free program is less than 100 K and available only for the Palm OS. Below, we selected a female age 60-64 with a high LDL (bad) <u>and</u> a high HDL (good) cholesterol (a common scenario in women) (Fig. 7.7). If she has no other risk factors to input, press calculate and you will see in figure 8 that her actual risk for coronary heart disease over the next 10 years is only 3%; whereas the average for her age is 12%. This program takes the guesswork out of determining the cardiac risk of elevated lipids and in addition offers links to other recommendations within this same program. The web site StatCoder also has calculators for the risk of stroke and atrial fibrillation. [8]

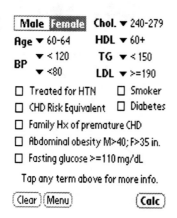

Figure 7.7. Patient risk factors

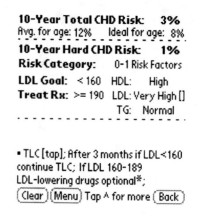

Figure 7.8. 10 Yr total CHD risk

3. **Depression.** A free program that uses a standard depression questionnaire and calculates the degree of depression.[8]

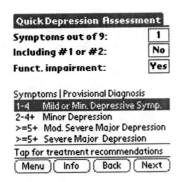

Figure 7.9. Depression assessment

4. **Growth Charts.** A free program that inputs the age, gender, height and weight and plots how the child compares to normative data.[8]

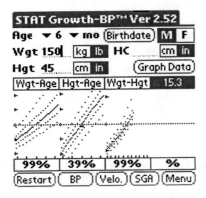

Figure 7.10. Growth Charts

5. **MedCalc .** A very popular Palm OS free program with over 75 commonly used formulas. You can customize the choices for only those formulas you use most often. It includes IV infusion rate calculators that should improve medication safety. [9]

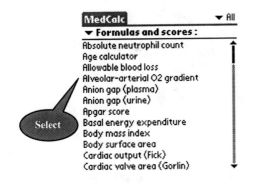 

Figure 7.11. Main screen   **Figure 7.12.** Calculating A-aO2 gradient

6. **Archimedes.** A free program for Windows OS and very similar to Medcalc.[10]

Figure 7.13. Anion gap

Figure 7.14. Basal Metabolic Rate

7. **ABG Pro.** A free Palm OS program that interprets arterial blood gases.[11]

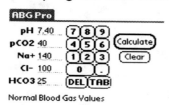

Figure 7.15. ABG Pro

8. **Preg Track.** Another free Palm OS program that tracks pregnancies and is also available in a desktop version.[11]

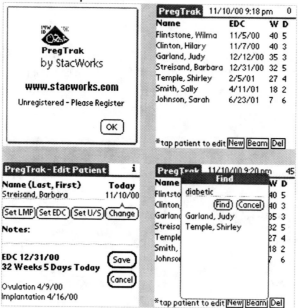

Figure 7.16. Preg Trak

9. **Evidence Based Medicine (EBM) Calculator.** A free program available for both the Palm OS and Windows OS systems. It calculates the relative risk reduction (RRR), the absolute risk reduction (ARR) and the number needed to treat (NNT) for random controlled trials (RCTs).[12] We will discuss this in more detail in the chapter on EBM.

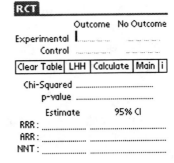

Figure 7.17. EBM Calculator   Figure 7.18. Random Controlled Trial

10. **Others.** We should note that Epocrates also offers the following free PDA calculators or "tools": drip rate calculator, cholesterol tool, MedMath, BMI tool, bone health tool, corticosteroid converter, depression assessment, cardiology essentials, heparin dosing, INR calculator, insulin calculator, temperature converter, topical steroid tool, tumor staging, hypertension tool, narcotic analgesic converter and GFR calculator.[13] ICU Math Medical calculator for the adult ICU uses 85 medical equations, including pulmonary, cardiology, BNP-CHF nomogram, pharmacokinetic dosing, renal, electrolyte, chemistry, nutrition, TPN, peri-operative risk, biostatistics, ACLS, Apache II, unit conversions and rules of thumb.[14]

PDA Textbooks

Trying to read a textbook on a small PDA screen will not appeal to everyone. Nevertheless, there are many textbooks available for use on the PDA. As a result of secure digital cards many textbooks can be stored on a memory card, thus saving space on the internal handheld memory. One of the most popular sites for mobile or e-textbooks is Skyscape.[10] They offer over 300 titles covering primary care and multiple specialties. The following are some of the most popular e-books requested: 5 Minute Clinical Consultant, Harrison's Manual of Medicine, The Washington Manual and The Harriet Lane Handbook.

Medical Software

Multiple medical PDA programs are available as freeware, shareware and fee based programs.

Shots 2006. A free program for Palm and Windows operating systems that is a good resource for childhood immunization schedules.[15]

US Preventive Services Task Force Tool. A free calculator program that provides the most recent national recommendations for screening and prevention, based on age, gender, etc. For Palm and Windows OS's.[16]

Pneumonia Severity Index. This free program calculates the severity and mortality of community acquired pneumonia. For Palm and Windows OS's.[17]

TB Treatment Guidelines. Originates from the Centers for Disease Control and is available for Palm OS only.[18]

There are simply too many PDA software programs to list. A March 2007 search for the word "medical" at the Palmgear website returned 570 software programs for the PDA.[19] Another excellent site for medical software programs is RNPalm (PDA Cortex).[20]

Drug Programs

Epocrates. While there are at least ten drug programs for the PDA, without question the most popular has been Epocrates, because it is intuitive, innovative and inexpensive. Although Figure 7.19 is not recent, it provides evidence of its tremendous popularity. According to a study at the Brigham and Woman's Hospital regarding the use of Epocrates: 82% thought it helped inform patients; 63% thought it reduced adverse drug events and 50% thought it reduced about one medication error per week. [21]

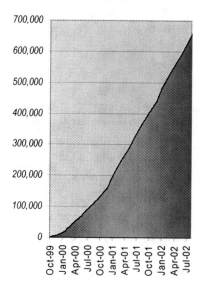

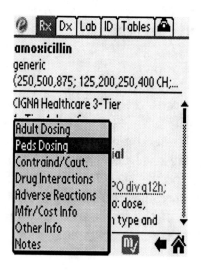

Figure 7.19. Popularity of Epocrates

Figure 7.20 Standard Menu

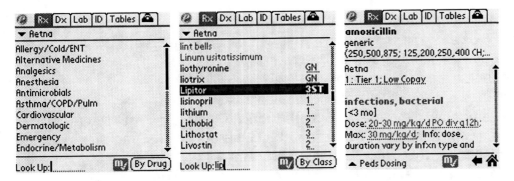

Figure 7.21. Note tabs   **Figure 7.22.** Note formulary   **Figure 7.23.** Drug indications in blue

As of March 2007, Epocrates is available for Palm and Windows OS's as the following programs:

1. *Epocrates Rx.* A free program with more limited drug content than fee based programs.
2. *Epocrates RxPro.* Provides the drug program plus an infectious disease (ID) program. $59.99 yearly
3. *Epocrates Essentials.* Includes a drug program, an infectious disease (ID) program, a lab test reference, Dx (5 Minute Clinical Consultant) and Sx that allows you to search symptom differential diagnoses. The drug program, the ID program and the 5 min Clinical Consultant are all integrated to make finding information easier. Essentials is $149 yearly and is available for Palm and Pocket PC OS's.
4. *Epocrates Essentials Deluxe.* Includes the Essentials package plus a medical dictionary (100,000 terms) and a coding tool (ICD-9 and CPT). Cost is $199 for one year, $299 for two years.
5. *Epocrates Dictionary.* Cost is $20 for one year, $35 for two years.
6. *Epocrates Coder.* Includes 20,000 ICD-9 and CPT codes. The cost is $70 for one year, $109 for two years.
7. *Online Epocrates.* There is now an Online Epocrates that offers the same features as the PDA program plus more than 100 calculations (MedCalc 3000), drug pictures, patient handouts and the ability to print or e-mail information. The program costs $59.99 yearly.
8. *Online Free Epocrates.* Since January 2006 there is a free online version of Epocrates. In October 2006 the option exists to add Epocrates Free Online as part of a Google home page. In December 2006 pill pictures, patient education and drug pricing/insurance information was added. Epocrates also now provides the code to add the search toolbar to any web site or Intranet at www.epocrates.com/medsearch
9. Epocrates Linx. Linx is an application program interface (API) that hyperlinks Epocrates to an electronic health record (EHR). With this API, drug questions can be answered without leaving the EHR.

Epocrates also has the following valuable features:
- The Medicare part D formulary that will be important with the new prescription benefits and e-prescribing
- Continuing medical information (CME)
- MedTools is a set of free medical calculators
- DxSx program that correlates more than 1200 diseases with symptoms. DXSx was developed by Massachusetts General Hospital and the developers of the 5 min Clinical Consultant
- Doc Alerts are medical alerts downloaded to your PDA when you synchronize
- Local formulary hosting available at an additional cost
- Black box warnings, therapeutic drug levels, monitoring parameters, and pregnancy and lactation warnings
- Metabolism, drug half life, mechanism of action and drug class [6]

Tarascon Pharmacopeia. It is similar to the popular pocket book and is available for Palm and Windows OS's. Simple to use with limited bells and whistles and costs only $27 yearly.[22]

Mobile Micromedex. Program is available for Palm or Windows OS's and free if an institution has a subscription to Micromedex; otherwise it costs $75. Overall, it is similar to Epocrates but in addition has toxicology tools.[23]

Thomson Clinical Xpert. Program is available free for Palm, Windows and BlackBerry OSs. It has a drug database, drug-drug interactions, toxicology, lab test resource, disease databases, calculators and medical news. [24]

Gold Standard's Clinical Pharmacology OnHand. Basic module has features similar to the other programs mentioned. Companion modules are: drug iDentifier, IV Alert and InFormulary. Future modules are Calculators and ACP Pier. Available for Palm or Windows OS. Basic module costs $99; addition of iDentifier and IV Alert adds $88.[25]

Physician Drug Handbook. Also available for Palm and Windows OS's and has all of the basic features already discussed. One of 32 pharmacy related textbooks for the PDA available at Skyscape, at a cost of $50.[10]

Infectious Disease Programs

Following calculators and drug programs, the next most popular downloadable category for the PDA is Infectious Disease (ID) programs. Like drug programs they provide quick answers to straightforward questions. According to one study, PDA Infectious Disease programs were found to be a good reference for a majority of general Internal Medicine admissions with infectious disease conditions.[26] These programs may not be able to answer highly complicated questions however. For an excellent review of these three ID PDA programs, we refer you to an article by Miller.[27]

Epocrates ID. This program can only be purchased as part of another Epocrates program. Sections are organized by bug, drug or location. It is intuitive and integrated with the Epocrates drug program. In Figure 7.26, for example, if you decide to use vancomycin it is hyperlinked back to the drug database which reduces the amount of taps of the stylus.[6]

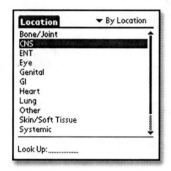

| **Figure 7.24.** Search by Location | **Figure 7.25.** Meningitis chosen | **Figure 7.26.** Therapy choice |

Sanford's Guide to Antimicrobial Therapy. The PDA version of the highly popular handbook is updated yearly. It is available for Palm, Windows and BlackBerry OS's with a annual cost of $27.50 [28]

John's Hopkins Antibiotic Guide. The Editor-in-Chief is the well-respected John Bartlett MD. The program is available as a desktop or PDA program with auto-updates. The program is organized by pathogen, antibiotic or disease and is free and available for the Palm OS only.[29]

Database Programs

Because PDAs retrieve and calculate data rapidly, a database program can be a valuable addition. Several excellent PDA database programs exist, but for the sake of brevity, only one program will be discussed.

HanDbase. This program creates a database on your desktop or on your PDA. Simply set up the fields, like name and beeper, with the characteristics you want the field to have and save with a name such as "Pagers July 2002". The database is created and saved on both your PC and your handheld. You can create small databases for inventories, lists of all of your usernames and passwords and so forth. It's easy to add a password to keep your information secure. The software is available for both Palm and Windows OSs. HanDbase is a relational database and not a flat file, so one database can be related to another database. The professional version comes with "Forms" that provides an attractive "front end" for data to be entered. See an example of forms in Figure 7.28 below. Over 2000 free databases are already created and available on the website; all you have to do is fill in the data. As an example, you can find a patient tracking or a wine inventory database. Databases can also synchronize to Microsoft Excel and Access. The cost for the professional version is $39.[30]

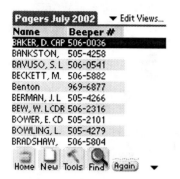
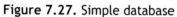

Figure 7.27. Simple database

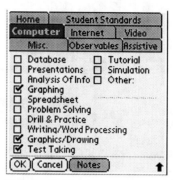

Figure 7.28. Form interface

Document Readers

Your PDA will not automatically read every software program or document. While searching sites such as Palmgear[19] you will occasionally see the comment that you need a document reader. There are numerous good products on the market but only one will be discussed.

iSilo. With this program you can read PDA documents that you download or create. Using iSilo, a program was written by one of the authors that converts a very lengthy guideline to prevent blood clots (DVTs) in hospitalized patients to the PDA format. Because the program was written in html it means that you can easily hyperlink one page to another. After a few screen taps, you can find the answer. Isilo is available for Palm and Windows OS's. The program will allow rich text, images and tables to be added at a cost of $20.[31]

Figure 7.29. Multiple docs   **Figure 7.30.** DVT Guidelines   **Figure 7.31.** Use of hyperlinking

IsiloX is a free desktop program that converts html documents, text documents and images so they can be seen on your PDA (in the isilo document reader). The above DVT program was converted with isiloX. This also means you can capture images, such as the coronary arteries on your PDA to show patients.[32]

Word documents on the PDA

Most Palm OS PDAs have the program "Documents to Go" that will convert standard Microsoft Word documents and Excel spreadsheets so they can be read on the PDA. If you own a Windows OS PDA, no conversion programs are needed to send Word documents or Excel spreadsheets to your PDA. The problem with placing long Word documents on the PDA is that they can be difficult to read due to the scrolling required. If you would like to create documents in html so pages can link to other pages (see figures 16-18) you can use isiloX and create a much more user friendly program. Their web site includes a forum and tutorials on converting text documents and web pages. [32]

Patient Tracking

Very few people are willing to manually input patient data into their PDA on a regular basis. It sounds good to say that you have all of your patient data on a PDA and you will beam the information to another physician at the end of your shift. The reality is that this is very time consuming. As mentioned, HanDbase has several free databases that could be used in this manner. Another program called Patient Tracker offers a PDA solution for $30.[33] Another similar product is Patient Keeper, with the exception that they also offer an enterprise solution that links to the hospital information system in order to retrieve lab data, capture charges, perform mobile dictation, send e-prescriptions and access a reference library.[34]

There are at least two other companies (Cogon and MercuryMD) that have been successful in integrating inpatient data to a PDA, smart phone or desktop format. These vendors are also able to post which patients you have in the hospital and their location as a rounding tool.[35-36]

Patient Billing

Mdeverywhere Everycharge. This billing program enables practices to enter charges into a Palm PDA or via the web on your desktop computer. A rules engine automatically identifies errors in charges to ensure accuracy and HIPAA compliance. EveryCharge "prompts the referring physicians, compares diagnoses and procedure code compatibility, includes payer-specific application rules, features an evaluation and management (E & M) coder and much more".[37]

Statcoder E & M. This is a simple program that calculates the correct evaluation and management (E & M) code by simply clicking on elements in several fields. (Figure 7.32). The program costs $75 for two years and includes a free trial. [8]

Figure 7.32. Statcoder E&M

Web Clipping

Avantgo. This site boasts over 1,000 potential web pages that can be downloaded to the PDA. The service is free but you are limited to 2MB of downloads. You can pay for the ability to download more content and you can pay more if you set up an enterprise program so others can download from your web site to their PDA. They have just added a beta RSS (really simple syndication) feed option for free. [38]

E-prescribing

E-prescribing will be discussed in detail under the chapter on e-prescribing.

Forms on the PDA

Pendragon Forms. With this program you can create customizable surveys. As an example, field workers could survey the safety status of a manufacturing plant and synchronize the information to the office PC at the end of the day. Data is uploaded as an Excel file that saves time and money by going paperless. Data can also be synchronized to Microsoft Access or a SQL server. The company also offers the ability to synchronize to a remote server so workers from multiple areas can send information to a head office. Cost for an individual license is $249 and is available for Palm and Window OS's. [39]

Wireless

There are currently four ways a PDA can be wireless:
1. PDA phones have features to access the Internet, e-mail and messaging.

2. 802.11x or WiFi will access the Internet via access points. Wireless is now part of higher end PDAs and you can also purchase a WiFi card that fits into the SDIO memory slot of earlier PDAs.
3. With Bluetooth you can connect wirelessly to computers, phones and printers if they are Bluetooth enabled. The maximal range is about 30 feet.
4. Infra-red beaming or synchronizing.

Other Uses

PDAs can be used for multiple other functions using the Palm or Windows operating systems.

1. Electrocardiograms (EKGs) and Pulmonary Function Tests (PFTs) can be performed using the PDA rather than a laptop or formal machine. The program requires the Windows OS.[40]
2. Bar code scanners are an option by buying a Symbol PDA or purchasing a bar code scanner that fits into the memory slot.[41]
3. The Global Positioning System (GPS) can utilize the PDA platform, as example, the iQue 3600 made by Garmin.[42]

Hardware Recommendations

It is difficult to make specific hardware recommendations due the differences in personal needs and the fact that companies produce new models and discontinue others at a rapid pace. It also depends on whether you think Palm will discontinue producing Palm OS PDAs in the near future. In general Palm OS PDAs are easier to use and less expensive. More medical software programs are written for Palm OS than Windows OS, but this is changing.

1. **Palm OS.** The Tungsten E2 is a good entry-level choice. The E2 model for $199 has the standard PDA functions with 32 MB of memory and Bluetooth. The new Palm TX has a larger screen, 100 MB of usable memory and is WiFi and Bluetooth enabled for only $299.[2]

2. **Windows OS.** Dell Axim PDAs tend to be less expensive than HP Ipaq models. The Dell Axim x51 model has a 415 MHz processor, 64 MB RAM, Bluetooth, but not WiFi, for $254. For a model with wireless and a faster processor the price is $399.[43]

BlackBerry

BlackBerry has added so many new features that they clearly should be considered "smart phones". One of the newest models (8800) includes e-mail, phone, web browsing, instant messaging, GPS, expandable memory (micro SD cards), maps, MP3 and video players. We would be remiss not to mention the sudden appearance of medical programs on this very popular platform. The following are some of the medical software that appeared in late 2006/early 2007 period:

- PEPID is a program of integrated treatment (2,700 topics) and drug content (6500 drugs) for emergency medicine, primary care and particularly for those in training programs
- Sanford Guide is a very popular infectious disease program, also available for the Palm and Windows operating systems
- Tarascon Pharmacopeia for drug lookups
- Lexi-comp ON-HAND that covers pharmacology, drug-drug interaction, lab and diagnostic tests, toxicology, dentistry and patient education [44]
- Enterprise Business (WLAN) solutions are available to incorporate nurse call functions, lab data, decision support, alerts, charge capture, e-prescribing, bar coding and RFID [45]
- Skyscape offers many book titles for the BlackBerry, traditionally used with the Palm and Windows operating systems. A secure digital memory card feature for most books is available [10]
- Thomson Clinical Xpert drug program is available free for the Blackberry, Palm and Windows OSs [24]

PDA Phones

More and more functions are appearing on cell phones to include complete PDA capabilities. Because the PDA is now integrated with a phone and access to the Internet new possibilities arise. As an example, real time monitoring of a variety of vital signs is possible. AIRSTRIP OB is a product that can transmit live fetal monitoring strips to an Obstetrician's cell phone. However, this program can only function if the OB unit has already purchased GE Centricity Prenatal.[46]

Several of the newest PDA phone models will be presented below.

1. **PPC6700 Sprint PDA Phone**
   a. Bluetooth and WiFi included
   b. MP3 player
   c. Video camera and player
   d. Slide out keyboard
   e. Mini-SD memory card
   f. Windows Mobile 5 OS
   g. 64 MB Ram
   h. 8+ days of standby time [47]

2. **Motorola Q**
   a. Very thin and weighs only 4.1 oz
   b. 64 MB Ram
   c. Mini-SD memory card

    d. Windows Mobile 5 OS
    e. 1.3 megapixel camera
    f. Video camera
    g. Dual CDMA channels
    h. Bluetooth enabled [48]

3. **Treo 700W**
    a. Windows Mobile 5 OS
    b. Carried by Verizon only
    c. Bluetooth enabled
    d. Wireless e-mail
    e. MP3 player
    f. 1.3 megapixel camera
    g. SD memory card
    h. 60 MB memory
    i. It offers average download speeds of 400-600 Kbps/sec
    j. Standby time of 15 days [49]

Limitations of mobile technology

This platform was never intended to replace the PC or laptop. In spite of improved speed, screen resolution and memory, several limitations persist:

1. Keyboard—simply too small and therefore slow to input significant information. You can purchase several different types of portable keyboards, including wireless, but it is one more thing to carry around. It will be interesting to see if voice recognition will become a viable option. The Treo 600/650 now has optional software for a few voice enabled commands.[50]
2. Small screen—several PDAs have larger screens but it is still difficult to scroll thru anything but short documents.
3. Security—passwords are rarely used with PDAs but should be if patient information resides on the PDA. 128-bit encryption is available[51] as are anti-virus programs.[52] This is important when PDAs synchronize to electronic health records or hospital information systems.
4. Frequently lost or stolen--in Chicago alone, taxi drivers recently reported that during a six-month period, 21,460 PDA's were accidentally left behind in their cabs. [51]

The Future of mobile technology

It is risky to try to predict the future, particularly when it comes to technology, but two trends will be mentioned.

1. The sale of Palm OS devices is stagnant overall. It has been stated that the sales of Treo phones (Palm) exceeds all other Palm OS device sales put together.[53] The fact that Palm now offers a PDA phone with its competition's operating system suggests that Palm might get out of the

PDA market. Several other companies including Sony have stopped producing PDAs.

2. The sale of smart phones worldwide has been phenomenal. Consumers will continue to insist on more features leading to more "smart phones" and PDA phones. It seems likely that the operating system of choice in the future will be Windows Mobile 5 in this country. Nokia, the heavy weight champion in Europe and elsewhere, uses the Symbian OS that has certain PDA functions, but at this point has limited medical applications available.[54]

PDA Resources

A Google search for "PDA software" in August 2006 returned over 5 million citations so we will only include a few of most commonly accessed sites

1. **Palmgear.** This company is one of the oldest and most comprehensive Palm PDA sites. It claims to have more than 28,000 software titles and lists 568 medical software programs consisting of freeware, shareware and programs for a fee. It also includes hardware, software, e-books, accessories and a "how to do it" section.[19]

2. **Handango.** Another complete PDA site that has been around for more than 5 years. A search on their site reveals more than 500 medical software downloads available. You are able to search by Palm or Windows operating systems.[55]

3. **American College of Physicians.** Includes a free 90 minute web presentation on "Practical Applications of Handheld Computers by Dr Peter Embi.[56]

4. **Skyscape.** It is known as the company with the most medical e-textbooks for the PDA, as it has more than 300 titles for nurses and physicians.[10]

5. **RNPalm (PDA Cortex).** Web site has an excellent selection of software programs for nurses and physicians.[20]

6. **Florida State University Medical School PDA site.** A very comprehensive site with tables of software programs, PDA web sites, "how to do it" and other valuable topics.[57]

7. **PDA TopSoft.** Web site has 286 medical software programs in almost every field, for every operating system.[58]

Conclusion

Mobile technology continues to improve at an amazing pace. Handheld units are being used for medical information, communication, monitoring and clinical decision support. In the not too distant future, they will be used for location, billing, data inputting and retrieval and functions we have yet to conceive. They should be considered a useful tool for clinicians and therefore part of an introductory course or textbook in Medical Informatics.

References

1. Everymac.com http://www.everymac.com/systems/apple/messagepad/stats/newton_mp_omp.html (Accessed February 1 2006)
2. PalmOne http://www.palm.com/us/products/palmpilot/ (Accessed August 30 2006)
3. Epocrates www.epocrates.com (Accessed March 7 2007)
4. PDA Medisoft http://www.pdamedisoft.com/BlackBerry (Accessed February 20 2007)
5. Criswell DF, Parchman ML Handheld Computer Use in the US Family Practice Residency Programs JAMIA 2002;9:80-86
6. Epocrates www.epocrates.com (Accessed March 7 2007)
7. Pediatrician's use of and attitudes about personal digital assistants Pediatrics 2004;113:238-24
8. Statcoder www.statcoder.com (Accessed February 1 2007)
9. Medcalc www.med-ia.ch/medcalc/download/html (Accessed February 1 2006)
10. Skyscape www.skyscape.com (Accessed February 1 2007)
11. ABG Pro http://www.stacworks.com/index.html (Accessed February 1 2006)
12. Center for Evidence Based Medicine http://www.cebm.utoronto.ca/practise/ca/statscal/ (Accessed February 5 2006)
13. Epocrates www.epocrates.com (Accessed March 7 2007)
14. FreewarePalm http://www.freewarepalm.com/medical/icumath.shtml (Accessed August 30 2006)
15. The Group on Immunization Education www.immunizationed.org (Accessed April 5 2006)
16. US Preventive Health Task Force http://pda.ahrq.gov/index.html (Accessed April 4 2006)
17. Agency for Healthcare Quality and Research http://pda.ahrq.gov/index.html15 (Accessed April 20 2006)
18. Centers for Disease Control http://www.cdc.gov/nchstp/tb/pubs/PDA_TBGuidelines/PDA_treatment_guidelines.htm (Accessed May 2 2006)
19. Palmgear www.palmgear.com (Accessed March 2 2007)
20. RNPalm www.rnpalm.com (Accessed May 20 2006)
21. Rothschild JM et al Clinician use of a palmtop drug reference guide JAMIA 2002;9:223-229
22. Tarascon https://www.tarascon.com/products/index.php?PID=9 (Accessed May 2 2006)
23. Mobile Micromedex http://www.micromedex.com/products/mobilemicromedex/ (Accessed May 5 2006)
24. Thomson Clinical Xpert http://www.pdr.net/pda-medical/ (Accessed March 2 2006) http://www.pdr.net/mobilepdr/mobilepdr.aspx (Accessed May 2 2006)
25. Gold Standard. http://www.clinicalpharmacologyonhand.com/marketing/about_cpoh.html (Accessed November 12 2006)

26. Miller SM, Beattie MM and Butt AA Personal Digital Assistant Infectious Diseases Applications for Health Care Professionals Clin Inf Dis 2003;36:1018-1029
27. Burdette SD, Herchline TE and Richardson WS Killing bugs at the bedside: a prospective hospital survey of how frequently personal digital assistants provide expert recommendations in the treatment of infectious diseases Ann Clin Micro and Antim. 2004;3 or http://www.ann-clinmicrob.com/content/3/1/22 (Accessed May 1 2006)
28. The Sanford Guide www.sanfordguide.com (Accessed May 1 2006)
29. Johns Hopkins Antibiotic Guide http://hopkins-abxguide.org/ (Accessed May 2 2006)
30. HanDbase www.ddhsoftware.com (Accessed June 2 2006)
31. isilo www.isilo.com (Accessed June 2 2006)
32. isiloX www.isiloX.com (Accessed June 2 2006)
33. Patient Tracker http://www.patienttracker.com/product_patienttracker.shtml (Accessed June 6 2006)
34. Patient Keeper http://www.patientkeeper.com/ (Accessed May 5 2006)
35. Cogon Systems www.cogonsystems.com (Accessed May 5 2006)
36. Mercury MD www.mercurymd.com (Accessed May 5 2006)
37. MDEverywhere www.mdeverywhere.com (Accessed April 5 2006)
38. Avantgo www.avantgo.com (Accessed May 5 2006)
39. Pendragon forms www.pendragon-software.com (Accessed May 19 2006)
40. Midmark http://www.midmarkdiagnostics.com/index.html (Accessed May 14 2006)
41. Palm barcoder http://www.ptshome.com/cs1504/cs1504palmbarcodekit.htm (Accessed May 2 2006)
42. Garmin GPS www.garmin.com (Accessed May 5 2006)
43. Dell Axim www.dell.com (Accessed June 5 2006)
44. BlackBerry https://www.blackberry.com/HealthcareSurvey/more.do;jsessionid=6GxvvLZG1eyoob-ZQRrWdg** (Accessed March 2 2007)
45. BlackBerry Enterprise http://na.blackberry.com/eng/solutions/industry/healthcare/ (Accessed March 2 2007)
46. AIRSTRIP OB www.airstripob.com (Accessed March 1 2007)
47. PPC6700 http://www.microsoft.com/windowsmobile/articles/ppc6700.mspx (Accessed May 5 2006)
48. Moto Q http://www.motorola.com/mdirect/q/index.html (Accessed May 6 2006)
49. Treo 700W http://web.palm.com/products/smartphones/treo700w/ (Accessed May 5 2006)
50. VoiceIT http://promo.palmgear.com/voiceit/ (Accessed May 5 2006)
51. Ultimaco http://americas.utimaco.com/products/pda/ (Accessed May 8 2006)
52. Avast http://www.avast.com/eng/download-avast-pda.html (Accessed May 21 2006)
53. Business Week http://www.businessweek.com/technology/content/may2005/tc20050519_1775_tc081.htm (Accessed May 5 2006)
54. The Register http://www.theregister.co.uk/2005/07/26/mobile_device_sales_q2_05/ (Accessed May 12 2006)
55. Handango www.handango.com (Accessed May 19 2006)
56. American College of Physicians www.acponline.org/pda (Accessed June 5 2006)
57. Florida State University http://med.fsu.edu/library/PDADocuments.asp (Accessed May 5 2006)
58. PDA TopSoft http://medical-software.pdatopsoft.com/?c=1 (Accessed May 18 2006)

Chapter 8: Evidence Based Medicine

Learning Objectives

After reading this chapter the reader should be able to:
- State the definition and origin of evidence based medicine
- Define the benefits and limitations of evidence based medicine
- Describe the evidence pyramid and levels of evidence
- State the process of using evidence based medicine to answer a medical question
- Compare and contrast the most important online and PDA evidence based medicine resources

Introduction

Some might ask why Evidence Based Medicine (EBM) is included in a textbook on Medical Informatics. The reason is that medical performance is based on quality and quality is based on the best available evidence. Clearly, information technology has the potential to improve decision making through online medical resources, electronic clinical practice guidelines, electronic health records (EHRs) with decision support, online literature searches, statistical analysis and online continuing medical education (CME). This chapter is devoted to how to find the best available evidence. Although one could argue that EBM is a buzz word like quality, in reality it means that clinicians should seek and apply the highest level of evidence available. According to the Center for Evidence Based Medicine, EBM can be defined as:

> "the conscientious, explicit and judicious use of current best evidence in making decisions about the care of individual patients" [1]

In *Crossing the Quality Chasm*, the Institute of Medicine (IOM) states:

> "Patients should receive care based on the best available scientific knowledge. Care should not vary illogically from clinician to clinician or from place to place" [2]

What the IOM is saying, is that every effort should be made to find the best answers and that these answers should be standardized and shared among clinicians. This is easier said than done because so many clinicians are independent practitioners with little allegiance to any one healthcare organization. It is true that many questions can not be answered by current evidence so clinicians may have to turn to subject experts. It is also true that the medical profession lacks the time and the tools to seek the best evidence.

One does not have to look very far to see how evidence changes recommendations:
- Bed rest is no longer recommended for low back pain; exercise is recommended instead [3]
- Bed rest is no longer recommended following a spinal tap (lumbar puncture); routine activity is recommended instead [4]

Until these older recommendations were challenged with high quality randomized controlled trials the medical profession had to rely on expert opinion, best guess or limited research studies.

The term EBM first appeared in 1992 in the Journal of the American Medical Association (JAMA).[5] One of the earliest proponents of EBM was Archie Cochrane, a British epidemiologist. Cochrane Centers and the international Cochrane Collaboration were named after him as a tribute to his early work. The Cochrane Collaboration consists of review groups, centers, fields, methods groups and a consumer network. Review groups, located in 13 countries, look at random controlled trials. As of 2005 they have completed about 2000+ systematic reviews, even though there have been 300,000 random controlled trials published.[6-7] The rigorous reviews are performed by volunteers, so efforts are slow. Dr David Sackett is another EBM pioneer who has been hugely influential at The Centre for Evidence Based Medicine in Oxford, England and McMaster University, Ontario, Canada. EBM has also been fostered at McMaster University by Dr Brian Haynes who is the Chairman of the Department of Clinical Epidemiology and Biostatistics and the editor of the American College of Physician's (ACP) Journal Club. Although EBM is popular in the United Kingdom and Canada it has received mixed reviews in the United States. The primary criticisms are that EBM tends to be a very labor intensive process and that in spite of the effort frequently no answer is found.

Random controlled trials started to appear in the 1960s. For the first time subjects who received a drug were compared with similar subjects who would receive another drug or placebo and the outcomes were evaluated. The studies became "double blind" meaning that both the investigators and the subjects did not know whether they received an active medication or a placebo. Until the 1980's evidence was summarized in review articles written by experts. Since the late 1980's more emphasis has been placed on improved study design, systematic reviews (to be explained later) and true outcomes research. It was no longer adequate to prove that a drug reduced blood pressure or cholesterol; it needed to demonstrate an actual change in outcome like reduced strokes or heart attacks.

Why is EBM Important?

Learning EBM is like climbing a mountain to gain a better view. You might not make it to the top and find the perfect answer but you will undoubtedly have a better vantage point than those who choose to stay at sea level. Reasons for studying EBM resources and tools include:

- Our current methods of keeping medically or educationally up to date do not work
- Translation of research into practice is often not successful
- Lack of time and the volume of published material result in information overload
- The pharmaceutical industry bombards clinicians and patients everyday; often with misleading or biased information
- Much of what we consider as the "standard of care" in every day practice has yet to be challenged and is probably wrong

Without proper EBM training we will not be able to appraise the best information resulting in poor clinical guidelines and wasted resources.

How have we traditionally gained medical knowledge?

1. Continuing Medical Education (CME). Traditional CME is desired by many clinicians but the evidence shows it to be highly ineffective. In general, busy clinicians are looking for a non-stressful evening away from their practice or hospital with food and drink provided.[8-9] Much of CME is provided free by pharmaceutical companies with their inherent biases. Better educational methods must be developed. A recent study demonstrated that online CME was at least comparable, if not superior to traditional CME.[10]
2. Clinical Practice Guidelines (CPGs). This will be covered in more detail in the next chapter. Unfortunately, just publishing CPGs does not in and of itself change how medicine is practiced.
3. Expert advice. Experts often approach a patient in a significantly different way compared to primary care clinicians because they deal with a highly selective patient population. Patients are often referred to specialists because the patient is not doing well and has failed treatment. For that reason, expert opinion needs to be evaluated along with EBM from the medical literature with the knowledge that expert recommendations may not be relevant to a primary care population. Expert opinion therefore should complement and not replace EBM.
4. Reading. It is clear that most clinicians are unable to keep up with medical journals published in their specialty. Most clinicians can only devote a few hours each week reading. All too often information comes from pharmaceutical representatives visiting the office. Moreover,

recent studies may contradict similar prior studies, leaving clinicians confused as to the best course.

What are the normal EBM steps towards answering a question?

1. You see a patient and realize you have a question
2. You next formulate a well constructed question. Here is the PICO method, developed by the National Library of Medicine, to formulate a question:
   - (P)atient or problem—how do you describe the patient group you are interested in? Elderly? Gender? Diabetic?
   - (I)ntervention—what is being introduced, a new drug or test?
   - (C)omparison—with another drug or placebo?
   - (O)utcome—what are you trying to measure? Mortality? Hospitalizations?

A web based PICO tool has been created by the National Library of Medicine to search Medline.[11] This tool could be placed on the desktop in the exam room or office. http://askmedline.nlm.nih.gov/ask/pico.php

3. Seek the best evidence for that question via an EBM resource or PubMed
4. Appraise that evidence using tools mentioned in this chapter
5. Apply the evidence to your patient.[12]

The most common types of clinical questions:
1. Therapy question. <u>This is the most common area for medical questions and the only one we will discuss in this chapter</u>
2. Prognosis question
3. Diagnosis question
4. Harm question
5. Cost question

The Evidence Pyramid

The pyramid in Figure 8.1 represents the different types of medical studies and their relative ranking. The starting point for research is often animal studies and the pinnacle of evidence is the meta-analysis. With each step up the pyramid our evidence is of higher quality associated with fewer articles published.[13] Although systematic reviews and meta-analyses are the most rigorous means to evaluate a medical question, they are expensive and labor intensive, hence more studies use a random controlled design.

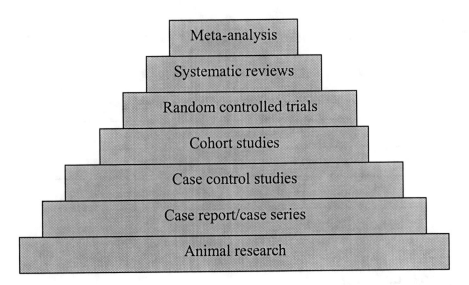

Figure 8.1. The Evidence Pyramid

Case reports/case series. Consist of collections of reports on the treatment of individual patients without control groups, as such has much less scientific significance.

Case control studies. Study patients with a specific condition and compare with people who do not. These types of studies are often less reliable than randomized controlled trials and cohort studies because showing a statistical relationship does not mean that one factor necessarily caused the other.

Cohort studies. Evaluate a large population and follow patients who have a specific condition or receive a particular treatment over time and compare them with another group that is similar but has not been affected by the condition being studied. Cohort studies are not as reliable as randomized controlled studies, since the two groups may differ in ways other than the variable under study.

Random controlled trials (RCTs). Subjects are randomly assigned to a treatment or no treatment or placebo group. RCTs include methods that reduce the potential for bias and that allow for comparison between intervention groups and control groups.

Systematic reviews. Multiple RCTs are evaluated to answer a specific question. Extensive literature searches are conducted to identify studies with sound methodology; a very time consuming process. The benefit is that multiple RCTs are analyzed, not just one study.

Meta-analyses. Takes the systematic review a step further by using statistical techniques to combine the results of several studies as if they were one large single study.[13]

We will be dealing exclusively with therapy questions so note that random controlled trials are the suggested study of choice (Table 8.1).[14]

Type of Question	Suggested best type of Study
Therapy	RCT > cohort > case control > case series
Diagnosis	Prospective, blind comparison to a gold standard
Harm	RCT > cohort > case control > case series
Prognosis	Cohort study > case control > case series
Cost	Economic analysis

Table 8.1. Best studies for questions asked

Levels of Evidence (LOE)

Unfortunately, there is no uniform level of evidence standard. The Center for Evidence Based Medicine offers an extremely detailed compilation of the levels of evidence that might seem a little overwhelming to someone new to EBM.[15] The following is a simpler version of the LOE:
- Level 1: Randomized Clinical Trials
- Level 2: Head to Head Trial or Systematic Review of Cohort Studies
- Level 3: Case-Control Studies
- Level 4: Case-series
- Level 5: Expert Opinion

Relative and Absolute Risk Reduction and the Number Needed to Treat

Overall, therapy trials are the most common area of research and ask questions such as, is drug A better than drug B or placebo? In order to determine what the true effect of a study is, it is important to understand the concepts of relative and absolute risk reduction and the number needed to treat.

Relative Risk Reduction (RRR) is the difference between the experimental event rate (EER), the therapy studied and the control event rate (CER), another drug or placebo, as a percentage of the control event rate. RRR = (EER-CER) divided by the CER. Here is an example:
- On Amazingstatin 5% (EER) of patients have a heart attack after 12 months of treatment. On Placebo 7.5% (CER) of patients have a heart attack over 12 months
- RRR = (7.5% - 5%) / 7.5% = 30%
- If you use RRR, Amazingstatin, on average, reduced the incidence of heart attacks by 30% compared to placebo over 12 months. Most of what you see written in the medical literature and the lay press will quote the RRR!

Absolute Risk Reduction (ARR) is the difference between the EER and the CER. In our example: ARR = 7.5% - 5% = 2.5%
- Using ARR we would say, on average, there is a 2.5% difference between the treatment group and the placebo group
- Do you think researchers, journal editors and drug companies will use the ARR or RRR?

Number Needed to Treat (NNT) is the absolute risk reduction (in percentage) divided into one hundred.[16] NNT = 100/2.5% = 40
- In our example it means that you have to treat (on average) 40 patients with Amazingstatin for 12 months to prevent one heart attack, compared to placebo
- This is a more meaningful way to report the results of therapy studies
- Unfortunately, very few studies offer NNT data, but it is very easy to calculate if you know the ARR. Nuovo, et al noted that NNT data was infrequently reported by five of the top medical journals in spite of being recommended [17]

Examples of using RRR, ARR and NNT. A full page article appeared in a December 2005 Washington Post newspaper touting the almost 50% reduction of strokes by a cholesterol lowering drug. This presented an opportunity to take a look at how drug companies usually advertise the benefits of their drugs. Firstly, in small print, you note that patients have to be diabetic with one other risk factor to see benefit. Secondly, there are no references. The statistics are derived from the CARDS Study published in the Lancet in Aug 2004.[18] Stroke was reported to occur in 2.8% in patients on a placebo and 1.5% in patients taking the drug Lipitor. The NNT is therefore 100/1.3 or 77. So, you had to treat 77 patients for an average of 3.9 years to prevent one stroke. This doesn't sound as good as "cuts the risk by nearly half". Now armed with these EBM tools, look further the next time you read about a miraculous drug effect.

Limitations of the Medical Literature and EBM

Because evidence is based on information published in the medical literature, it is important to point out some of the limitations researchers and clinicians must deal with on a regular basis:
- Low yield of clinically useful articles in general [19]
- Conflicts of interest as a result of pharmaceutical company influence [20-21]
- Up to 16% of well publicized articles are contradicted in subsequent studies [22]
- Peer reviewers are "unpaid, anonymous and unaccountable" so it is often not known who reviewed an article and how rigorous the review was [23]

- Many medical studies are poorly designed [24]
  - The recruitment process was not described [25]
  - Many randomized trials (particularly drug company sponsored trials) are stopped prematurely due to early benefit. It is possible that benefit would not be seen in a longer study [26]
  - Inadequate power (size) to make accurate conclusions. In other words, not enough subjects were studied [27]

Limitations of EBM

In spite of the fact that EBM is considered a highly academic process towards gaining medical truth, numerous problems exist:
- Different rating systems by various medical organizations
- Different conclusions by experts evaluating the same study
- Time intensive exercise to evaluate existing evidence
- Systematic reviews are limited in the topics reviewed and are time intensive to complete (6-24 months). Often the conclusion is that current evidence is weak and further high quality studies are necessary
- Random controlled trials are expensive. Drug companies tend to fund only studies that help a current non-generic drug they would like to promote
- Studies with negative results are not always published (publication bias)
- Results may not be applicable to every patient population
- Some view EBM as "Cookbook Medicine" [28]
- There is not good evidence that teaching EBM changes behavior [29]

Other Approaches

EBM has had both strong advocates and skeptics since its inception. One of its strongest proponents, Dr David Sackett published his experience with an "Evidence Cart" on inpatient rounds in 1998. The cart contained numerous EBM references but was so bulky that it could not be taken into patient rooms. [30] Since that article, multiple, more convenient EBM solutions exist. While there are those EBM advocates who would suggest we use solely EBM resources, many others feel that EBM "may have set standards that are untenable for practicing physicians". [31-32]

Dr Frank Davidoff believes that most clinicians are too busy to perform literature searches for the best evidence. He believes that we need "Informationists", who are experts at retrieving information. [33] To date, only clinical medical librarians (CMLs) have the formal training to take on this role. At large academic centers CMLs join the medical team on inpatient rounds and attach pertinent and filtered articles to the chart. As an example, Vanderbilt's

Eskind Library has a Clinical Informatics Consult Service. [34-35] The obvious drawback is that CMLs are only available at large medical centers and are unlikely to research outpatient questions.

According to Slawson and Shaughnessy you must become an "information master" to sort through the "information jungle". They define the usefulness of medical information as:

$$\text{Usefulness} = \frac{\textit{Validity} \times \textit{Relevance}}{\textit{Work}} \text{ [36]}$$

Only the clinician can determine if the article is relevant to his/her patient population and if the work to retrieve the information is worthwhile. Slawson and Shaughnessy also developed the notion of looking for "patient oriented evidence that matters" (POEM) and not "disease oriented evidence that matters" (DOEM). POEMS look at mortality, morbidity and quality of life whereas DOEMS tend to look at laboratory or experimental results. They point out that it is more important to know that a drug reduces heart attacks or deaths from heart attacks, rather than just reducing cholesterol levels (DOEM). This school of thought also recommends that you not read medical articles blindly each week but should instead learn how to search for patient specific answers using EBM resources. [37] This also implies that you are highly motivated to pursue an answer and have the appropriate training.

EBM Resources

There are many first-rate online medical resources that provide EBM type answers. They are all well referenced, current and written by subject experts. Several include the level of evidence (LOE). Examples would include UpToDate, First Consult, InfoRetriever, DynaMed, ACP Medicine, ACP-PIER and E-medicine. For the EBM purist, the following are considered traditional or classic EBM resources:

- Clinical Evidence [38]
  - British Medical Journal product with two issues per year
  - New drug safety alert section
  - New "latest research" results section
  - Evidence is oriented towards patient outcomes (POEMS)
  - Very evidence based with single page summaries and links to national guidelines
  - Available in paperback (Concise), CD-ROM, online or PDA format
- Cochrane Library [39] (fee based) consists of :
  - Database of systematic reviews. Each review answers a clinical question
  - Database of review abstracts of effectiveness (DARE)
    - Controlled Trials Register
  - Methodology reviews

- o   Methodology register
- Cochrane Review [40]
  - o   Part of the Cochrane Collaboration
  - o   Reviews can be accessed for a fee but abstracts are free. A search for low back pain, as an example, returned 44 reviews (abstracts)
- BMJ Updates [41]
  - o   Since 2002 BMJ Updates has been filtering all of the major medical literature. Articles are not posted until they has been reviewed for newsworthiness and relevance; not strict EBM guidelines
  - o   You can go to their site and do a search or you can choose to have article abstracts e-mailed to you on a regular basis
  - o   These same updates are available through www.Medscape.com
- ACP Journal Club [42]
  - o   Bimonthly journal that can be accessed from OVID or free if a member of the American College of Physicians (ACP)
  - o   Over 100 journals are reviewed but very few articles make the cut: in 1992 only 13% of articles from the NEJM made the Journal Club, all other journals were much lower
- Practical Pointers for Primary Care [43]
  - o   Free online review of articles written in the New England Journal of Medicine, Journal of the American Medical Journal, British Medical Journal, the Lancet, the Annals of Internal Medicine and the Archives of Internal Medicine
  - o   Dissects the study and makes summary comments, helpful to clinicians
- PDA EBM Resources
  - o   Centre for EBM www.cebm.utoronto.ca
  - o   Duke Medical Center Library www.mclibrary.duke.edu/training/pdaformat/
  - o   EBM 2 go http://www.ebm2go.com
- Others
  - o   TRIP Database www.tripdatabase.com/
  - o   OVID http://gateway.ovid.com Ability to search Cochrane Database of Systematic Reviews, DARE, ACP Journal Club and Cochrane Controlled Trials Register at the same time
  - o   SUMSearch http://sumsearch.uthscsa.edu. Free site that searches Medline, National Guideline clearing house and DARE
  - o   Bandolier www.jr2.ox.ac.uk/bandolier Free EBM journal; used mainly by primary care docs in England. Provides simple summaries with NNTs
  - o   Evidence Based Medicine Resources http://denison.uchsc.edu/evidence_based.html Denison Memorial Library

Conclusion

Knowledge of evidence based medicine is important if you are involved with patient care or quality of care issues. Rapid access to a variety of online EBM resources has changed how we practice medicine. In spite of its shortcomings, an evidence based approach helps healthcare workers find the best possible answers. Busy clinicians are likely to choose commercial high quality resources, while academic clinicians are likely to select true EBM resources. Ultimately, EBM tools and resources will be integrated into all electronic health records.

References

1. Evidence Based Medicine: What it is, what it isn't. http://www.cebm.net/ebm_is_isnt.asp (Accessed September 3 2005)
2. Crossing the Quality Chasm: A new health system for the 21th century (2001) The National Academies Press http://www.nap.edu/books/0309072808/html/
3. MedlinePlus http://www.nlm.nih.gov/medlineplus/ency/article/003108.htm (Accessed September 3 2006)
4. Teece I, Crawford I. Bed rest after spinal puncture. BMJ http://emj.bmjjournals.com/cgi/content/full/19/5/432 (Accessed Aug 24 2006)
5. Guyatt et al Evidence-based medicine: a new approach to teaching the practice of medicine JAMA 1992;268:2420-5
6. Evidence Based Medicine . Wikipedia. http://en.wikipedia.org/wiki/Evidence_based-medicine (Accessed September 5 2005)
7. Levin A The Cochrane Collaboration Ann of Int Med 2001;135:309-312
8. Davis D A, et al. Changing physician performance. A systematic review of the effect of continuing medical education strategies. JAMA 1995; 274: 700-1.
9. Sibley J C, A randomized trial of continuing medical education. N Engl J Med 1982; 306: 511-5.
10. Fordis M et al. Comparison of the Instructional Efficacy of Internet-Based CME with Live Interactive CME Workshops. JAMA 2005;294:1043-1051
11. National Library of Medicine PICO http://askmedline.nlm.nih.gov/ask/pico.php (Accessed September 7 2005)
12. Centre for Evidence Based Medicine http://www.cebm.net/learning_ebm.asp (Accessed September 6 2005)
13. Haynes RB Of studies, syntheses, synopses and systems: the "4S evolution of services for finding the best evidence". ACP J Club 2001;134: A11-13
14. The well built clinical question. University of North Carolina Library http://www.hsl.unc.edu/Services/Tutorials/EBM/Supplements/QuestionSupplement.ht m (Accessed September 20 2005)
15. Evidence Based Medicine: What it is, what it isn't. http://www.cebm.net/ebm_is_isnt.asp (Accessed September 3 2005)
16. Henley E Understanding the Risks of Medical Interventions Fam Pract Man May 2000;59-60
17. Nouvo J, Melnikow J, Chang D Reporting the Number Needed to Treat and Absolute Risk Reduction in Randomized Controlled Trials JAMA 2002;287:2813-2814
18. Collaborative Atorvastatin Diabetes Study (CARDS) Lancet 2004;364:685-96
19. Haynes RB Where's the Meat in Clinical Journals? ACP Journal Club Nov/Dec 1993: A-22-23
20. Friedman LS, Richter ED Relationship between conflicts of interest and research results J of Gen Int Med 2004;19:51-56
21. Drazen FM Financial Associations with Authors NEJM 2002;346:1901-2

22. Ioannidis JPA, Contradicted and Initially Stronger Effects in Highly Cited Clinical Research JAMA 2005;294:218-228

23. Kranish M Flaws are found in validating medical studies The Boston Globe August 15 2005

24. Altman DG Poor Quality Medical Research: What can journals do? JAMA 2002;287:2765-2767

25. Gross CP et al Reporting the Recruitment Process in Clinical Trials: Who are these Patients and how did they get there? Ann of Int Med 2002;137:10-16

26. Montori VM et al Randomized Trials Stopped Early for Benefit, a systematic review JAMA 2005;294:2203-2209

27. Moher D, Dulgerg CS, Wells GA Statistical Power, sample size and their reporting in randomized controlled trials JAMA 1994;22:1220-1224

28. Straus SE, McAlister FA Evidence Based Medicine: a commentary on common criticisms Can Med Assoc J 2000;163:837-841

29. Dobbie AE et al What Evidence Supports Teaching Evidence Based Medicine? Acad Med 2000;75:1184-1185

30. Sackett DL, Staus SE Finding and Applying Evidence During Clinical Rounds: The "Evidence Cart" JAMA 1998;280:1336-1338

31. Grandage, K. et al. When less is more: a practical approach to searching for evidence-based answers. J Med Libr Assoc 90(3) July 2002

32. Schilling LM et al Resident's Patient Specific Clinical Questions: Opportunities for Evidence Based Learning Acad Med 2005;80:51-56

33. Davidoff F, Florance V The Informationist: A New Health Profession? Ann of Int Med 2000;132:996-999

34. Giuse NB et al Clinical medical librarianship: the Vanderbilt experience Bull Med Libr Assoc 1998;86:412-416

35. Westberg EE, Randolph AM The Basis for Using the Internet to Support the Information Needs of Primary Care JAMIA 1999;6:6-25

36. Slawson DC, Shaughnessy AF, Bennett JH Becoming a Medical Information Master: Feeling Good About Not Knowing Everything J of Fam Pract 1994;38:505-513

37. Shaughnessy AF, Slawson DC and Bennett JH Becoming an Information Master: A Guidebook to the Medical Information Jungle J of Fam Pract 1994;39:489-499

38. Clinical Evidence www.clinicalevidence.com (Accessed January 23 2007)

39. Cochrane Library http://www3.interscience.wiley.com/cgi-bin/mrwhome/106568753/HELP_Cochrane.html (Accessed January 23 2007)

40. Cochrane Review http://www.cochrane.org/reviews/index.htm (Accessed September 18 2005)

41. Bmjupdates http://bmjupdates.mcmaster.ca (Accessed January 25 2007)

42. ACP Journal Club http://www.acpjc.org/ (Accessed January 26 2007)

43. Practical Pointers for Primary Care www.practicalpointers.org (Accessed January 26 2007)

Chapter 9: Clinical Practice Guidelines

Learning Objectives

After reading this chapter the reader should be able to:
- Define the utility of clinical practice guidelines
- Describe the interrelationship between clinical practice guidelines, evidence based medicine, electronic health records and pay for performance
- Define the processes required to write and implement a clinical practice guideline
- Compare and contrast the potential benefits and obstacles of clinical practice guidelines
- Describe clinical practice guidelines in an electronic format
- List the most significant clinical practice guideline resources

Introduction

The following is a definition of Clinical Practice Guidelines (CPGs):

> *"A set of systematically developed statements or recommendations designed to assist practitioner and patient decisions about appropriate health care for specific clinical circumstances"* [1]

Information technology assists CPGs by expediting the search for the best evidence and linking the information to EHRs, web pages and Intranets for easy access. Clinical practice guidelines (CPGs) take the very best evidence based medical information and formulates a game plan to treat a specific disease or condition. Many medical organizations use CPGs with the intent to improve quality of care, patient safety and/or reduce costs. Two areas CPGs may be potentially beneficial in include:
- Use in "pay for performance" programs
- Use in disease management programs:
  - 83% of Medicare beneficiaries have at least one chronic condition and 68% of Medicare's budget is devoted to the 23% who have 5 or more chronic conditions [2]
  - There is some evidence that guidelines that address multiple co-morbidities (associated illnesses) actually do work. As an example, in one study of diabetics, there was a 50% decrease in cardiovascular and microvascular complications with intensive treatment of multiple risk factors [3]

In spite of evidence to suggest benefit, several studies have shown poor CPG compliance by patients and physicians. The well publicized 2003 RAND study in the New England Journal of Medicine demonstrated that "overall, patients received 54% of recommended care".[4-5] In another study of guidelines at a major teaching hospital there was overuse of statin therapy (cholesterol lowering drugs). Overuse occurred in 69% of cases of primary prevention (to prevent a disease) and 47% overuse in secondary prevention (to prevent disease reoccurrence), compared to national recommendations.[6]

It should be emphasized that creating or importing a guideline is the easy part. Implementing CPGs and achieving buy-in by all healthcare workers (particularly physicians) is the hard part.

Where should hospitals or individuals start? Examples:
- High cost conditions: heart failure
- High volume conditions: diabetes
- Preventable admissions: asthmatics
- Where you suspect you have variation in care compared to national recommendations: deep vein thrombophlebitis (DVT) prevention
- High litigation areas: failure to diagnose or treat
- Pressing Patient Safety areas: intravenous (IV) drug monitoring

Attributes of a Good Guideline
- Evidence based
- Updated yearly
- Good medical references
- Level of evidence given
- Simple summary or algorithm that is easy to understand
- Available in multiple formats (print, online, PDA, etc.) in multiple locations
- Compatibility with existing practices
- Simplifies, not complicates decision making [7]

Barriers to CPGs
- Practice setting: inadequate incentives, inadequate time and fear of liability
- Contrary opinions: local experts don't agree or different message from drug companies
- Knowledge and attitudes: there is a lack of confidence to either not perform a test (malpractice concern) or to order a new treatment (don't know enough yet). Information overload is always a problem [8]
- CPGs are too long, impractical or confusing. One study of Family Physicians stated CPGs should be no longer than 2 pages.[9-10] Most national CPGs are 50-150 pages long and don't always include a summary of recommendations
- Where do you post the CPG and in what format?

- If you don't report local data you will have less buy-in. Physicians tend to respond to data from their hospital or clinic
- No uniform level of evidence (LOE) rating system
- No available local champions
- Many CPG's posted on National Guideline Clearinghouse addressing the same disease/topic making standardization and selection of best guideline more challenging[11]
- Excessive influence by drug companies: Survey of 192 authors of 44 CPGs 1991-1999
  - 87% had some tie to drug companies
  - 58% received financial support
  - 59% represented drugs mentioned in the CPG
  - 55% of respondents with ties to drug companies said they didn't believe they had to disclose involvement [12]

The Strategy
- Leadership support is crucial
- Use process improvement tools such as the Plan-Do-Study-Act model
- Identify gaps in knowledge between national recommendations and local practice
- Locate a guideline champion who is a well respected clinician expert.[13] A champion acts as an advocate for implementation based on his/her support of a new guideline.
- Other potential team members
  - Clinicians selected based on the nature of the CPG
  - Administrative or support staff
  - Quality Management staff
- Develop action plans
- Educate all staff involved with CPG, not just clinicians
- Pilot implementation
- Provide frequent feedback to clinicians and other staff regarding results

Examples of Clinical Practice Guidelines

The following examples are courtesy of the Naval Hospital Pensacola. The CPG in figure 9.1 was written for the treatment of uncomplicated bladder infections (cystitis) in women. The goal was to use less expensive antibiotics and for fewer days. The protocol or algorithm can be administered by a triage nurse when a patient telephones or walks in.

Figure 9.2 demonstrates that the use of the first line drug (sulfa family) increased after the start of the CPG, whereas second line drug use decreased. Success of this program was based on educating all members of the healthcare team and reporting the results at medical staff meetings and other venues. It was also aided by an easy to follow guideline and full support by the nursing staff.

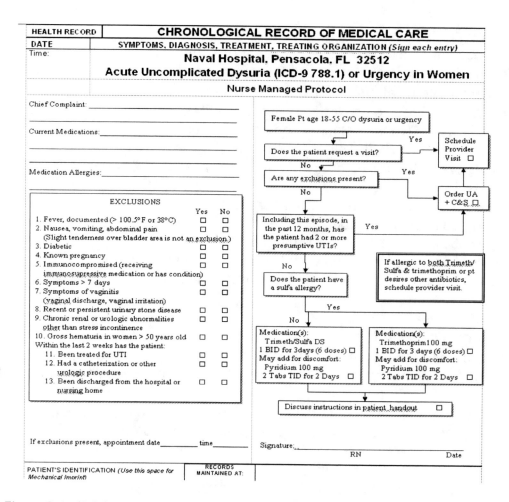

Figure 9.1. CPG for uncomplicated dysuria or urgency in women

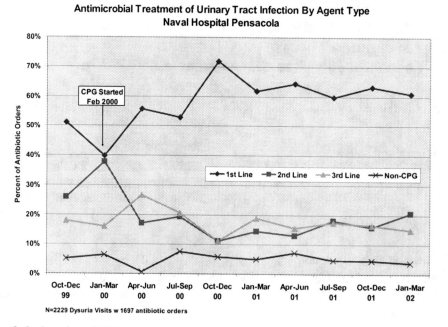

Figure 9.2. Results of CPG implementation

Figure 9.2 demonstrates that the use of the first line drug (sulfa family) increased after the start of the CPG, whereas second line drug use decreased. Success of this program was based on educating all members of the healthcare team and reporting the results at medical staff meetings and other venues. It was also aided by an easy to follow guideline and full support by the nursing staff.

Clinical Practice Guidelines in electronic formats

CPGs have been traditionally paper based and often accompanied by a flow diagram or algorithm. With time, more are being electronic and posted on the Internet or Intranet for easy access. Others are available for download to a personal digital assistant. PDAs function well in this area as each step in an algorithm is simply a tap on the screen. In addition to commercially hosted PDA programs, CPGs can be created on the PDA using isiloX and simple hyperlinks (see chapter on Mobile Technology). In figures 9.3 and 9.4 PDA programs are shown that are based on national guidelines for cardiac risk and cardiac clearance. Figure 9.3 depicts a calculator that determines the 10 year risk of heart disease based on your serum cholesterol and other risk factors. A cardiac clearance program that determines whether a patient needs further cardiac testing prior to an operation is noted in figure 9.4.[14] Many excellent PDA guidelines exist and these will be listed later in this chapter.

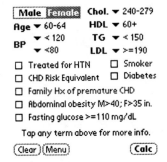

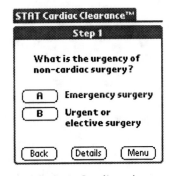

Figure 9.3. 10 year risk of heart disease　　Figure 9.4. Cardiac clearance

Another simple technique to post a CPG in a user friendly manner is to write it as a Word document but save it as a web page (html). Hyper linking to other pages or simply "book marking" to information located further down on the same page is an advantage. This allows a CPG to be written on a single document that is easy to e-mail or post. The 2004 American College of Chest Physicians (ACCP) Guideline to prevent venous thromboembolism (blood clots in legs and lungs) was converted to this format with general recommendations (Figure 9.5) hyperlinked to specific recommendations to make the guideline compact and easy to navigate.

PREVENTION OF VENOUS THROMBOEMBOLISM   Naval Hospital Pensacola   September 2004

General Surgery   Gynecologic Surgery   Urologic Surgery   Orthopedic Surgery   Neurosurgery and Trauma

Medical Conditions   Reference   Risk Assessment

General recommendations:

Every hospital should develop a written policy or other formal strategy for preventing thromboembolic complications, especially for high-risk patients.

We do **not** recommend prophylactic use of aspirin as a sole prevention strategy, because the measures listed in the tables below are more efficacious (**grade 1A**).

We recommend anticoagulant prophylaxis or therapy should be used with **caution** in all cases involving either spinal puncture or placement of epidural catheters for regional anesthesia or analgesia (**grade 1C+**)

Level of Evidence:

Grade of recommendation	Implications
1A	Strong. Applies to most patients in most circumstances
1B	Strong. Likely to apply to most patients
1C+	Strong. Can apply to most patients in most circumstances
1C	Intermediate. May change when stronger evidence is available

Glossary:

ES=elastic stockings   **IPC**=intermittent pneumatic compression

LDUH=low dose unfractionated heparin   **LMWH**=low molecular weight heparin

GENERAL SURGERY

RISK GROUP	RECOMMENDED PROPHYLAXIS
Low risk (Minor procedure in patients< 40 yr, with no additional risk factors)	Early ambulation only (**Grade 1C+**)
Moderate risk (Non-major surgery in patients 40 to 60 yr or have additional risk factors or those who are undergoing major surgery and are < 40 yrs with no additional risk factors)	LDUH 5,000 units BID or LMWH once daily (**Grade** I A)
Higher risk (Non-major surgery in patients > 60 yr, or with additional risk factors or major surgery in patients > 40 yrs or with additional risk factors)	LDUH 5,000 units TID or LMWH > 3,400 U daily (**Grade** IA)
High risk with multiple risk factors	LDUH 5,000 units TID or LMWH >3,400 U daily plus ES and/or IPC (**Grade** IC+)
In selected high risk patients including those who have undergone major cancer surgery	Discharge on LMWH (**Grade** 2A)
Higher risk patients with greater-than-usual risk for bleeding	ES or IPC at least initially until bleeding risk decreases (**Grade** IA)

Figure 9.5. DVT prevention CPG (courtesy Naval Hospital Pensacola)

Only a minority of electronic health records have embedded CPGs but there is definite interest in providing local or national CPGs at the point of care. CPGs embedded in the EHR are clearly a form of decision support. They can be linked to the diagnosis or the order entry process. In addition, they can be standalone resources available by clicking, for example, an "info-button". The obvious goal of this clinical decision support is to provide treatment reminders for disease states that may include the use of more cost effective drugs. Institutions such as Vanderbilt University have integrated more than 750 CPGs into their EHR by linking the CPGs to ICD-9 codes.[15] The results of embedded CPGs appears to mixed. In a study by Durieux using computerized decision support reminders, Orthopedic surgeons showed improved compliance to guidelines to prevent deep vein thrombophlebitis [16] On the other hand, three studies by Tierney, failed to demonstrate improved compliance to guidelines using computer reminders for hypertension, heart disease and asthma. [17-19]

There are other ways to use electronic tools to promulgate CPGs. In an interesting paper by Javitt, primary care clinicians were sent reminders on outpatient treatment guidelines based only on claims data. Outliers were located by using a rules engine (Care Engine) to compare a patient's care with national guidelines. They were able to show a decrease in hospitalizations and cost as a result of alerts that notified physicians by phone, fax or letter. This demonstrates one additional means of changing physician behavior using CPGs and information technology, not linked to the electronic health record. [20]

CPG Resources

National Guideline Clearinghouse (NGC) (www.guideline.gov).
NGC is an initiative of the Agency for Healthcare Research and Quality (AHRQ), U.S. Department of Health and Human Services. It is the largest and most comprehensive CPG resource with over 2000 guideline summaries and links to national and international medical organizations with CPG-related information. Features include:
- CPG browsing options such as by disease, treatment, organizations
- NGC templated CPG attributes summarized as well as links to full text guidelines when available
- Comparison tools and synthesis of CPG's when more than one exists about a topic.
- Forum for discussion of guidelines
- Annotated bibliography
- Patient resources
- Evidence-based practice center (EPC) reports and technology assessments
- For handheld computers/PDA's (Table 9.1):
    o Option to download NGC guideline summaries directly to Pocket PC OS or via document reader for Palm OS.
    o Links to organizations with CPG's downloadable to a PDA. [11]

PDA Downloadable Clinical Practice Guidelines
National Guideline Clearinghouse (NGC) (www.guideline.gov/resources/pda.aspx) • NGC guideline summaries
Available from Apprisor (www.apprisor.com/dlselect.cfm) • Free document reader for Pocket PC and Palm OS • American Academy of Family Physicians (AAFP) Summary of Recommendations for Clinical Preventive Services 2007 • American College of Cardiology (ACC)/American Heart Association (AHA) Joint Guidelines • American College of Chest Physicians (ACCP) Guidelines • American College of Physicians (ACP) Guidelines • National Heart, Lung and Blood Institute JNC VII on High Blood Pressure
American College of Radiology Appropriateness Criteria for imaging decisions (www.acr.org/s_acr/sec.asp?CID=1845&DID=16050) • PDA version requires annual subscription • Pdf version free download
American Diabetes Association (ADA) clinical practice guidelines 2006 (www.diabetes.org/for-health-professionals-and-scientists/cpr-pda.jsp)

Table 9.1 PDA Downloadable Clinical Practice Guidelines

Conclusion

The jury is out regarding the impact of CPGs on physician behavior or patient outcomes. Busy clinicians are slow to accept new information, including CPGs. Whether embedding CPGs into EHRs will result in significant changes in behavior that will consistently result in improved quality, patient safety or cost savings remains to be seen. It is also unknown if linking CPGs to better reimbursement (pay for performance) will result in a higher level of acceptance. While we are determining how to optimally improve healthcare with CPGs, most authorities agree that CPGs need to be concise, practical and accessible at the point of care.

References

1. Program of excellence www.excellence.dxu.com/Glossary.htm (Accessed April 1 2006)
2. O'Connor P Adding Value to Evidence Based Clinical Guidelines JAMA 2005;294:741-743
3. Gaede P Multifactorial intervention and cardiovascular disease in patients with type 2 diabetes NEJM 2003;348:383-393
4. McGlynn E Quality of Health Care Delivered to Adults in the US RAND Health Study NEJM Jun 26 2003
5. Crossing the Quality Chasm: A new Health System for the 21th century 2001. IOM. http://darwin.nap.edu/books/0309072808/html/227.html (Accessed March 5 2006)
6. Abookire SA, Karson AS, Fiskio J, Bates DW Use and monitoring of "statin" lipid-lowering drugs compared with guidelines Arch Int Med 2001;161:2626-7
7. Oxman A, Flottorp S. An overview of strategies to promote implementation of evidence based health care. In: Silagy C, Haines A, eds Evidence based practice in primary care, 2nd ed. London: BMJ books 2001

8. Grol R, Grimshaw J From Best evidence to best practice: effective implementation of change in patient's care Lancet 2003;362:1225-30
9. Wolff M, Bower DJ, Marabella AM, Casanova JE US Family Physicians experiences with practice guidelines. Fam Med 1998;30:117-121
10. Zielstorff RD Online Practice Guidelines JAMIA 1998;5:227-236
11. National Guideline Clearinghouse www.guideline.gov (Accessed May 5 2007)
12. Choudry NK et al Relationships between authors of clinical practice guidelines and the pharmaceutical industry JAMA 2002;287:612-7
13. Stross JK– The educationally influential physician – Journal of Continuing Education Health Professionals 1996; 16: 167-172)
14. Cardiac Clearance www.statcoder.com (Accessed March 2 2007)
15. Giuse N et al Evolution of a Mature Clinical Informationist Model JAIMA 2005;12:249-255
16. Durieux P et al A Clinical Decision Support System for Prevention of Venous Thromboembolism: Effect on Physician Behavior JAMA 2000;283:2816-2821
17. Tierney WM et al Effects of Computerized Guidelines for Managing Heart Disease in Primary Care J Gen Int Med 2003;18:967-976
18. Murray et al Failure of computerized treatment suggestions to improve health outcomes of outpatients with uncomplicated hypertension: results of a randomized controlled trial Pharmacotherapy 2004;3:324-37
19. Tierney et al Can Computer Generated Evidence Based Care Suggestions Enhance Evidence Based Management of Asthma and Chronic Obstructive Pulmonary Disease? A Randomized Controlled Trial Health Serv Res 2005;40:477-97
20. Javitt JC et al Using a Claims Data Based Sentinel System to Improve Compliance with Clinical Guidelines: Results of a Randomized Prospective Study Amer J of Man Care 2005;11:93-102

Chapter 10: Disease Management and Disease Registries

Learning Objectives

After reading this chapter the reader should be able to:
- Define the role of disease management in chronic disease
- Describe the need for rapid retrieval of patient and population statistics to manage patients with chronic diseases
- Compare and contrast the various disease registry formats to include integration with electronic health records
- Describe the interrelationships between disease registries, evidence based medicine and pay for performance

Introduction

According to Epstein, Disease Management (DM) is:

> *"a systematic population based approach to identify persons at risk, intervene with a specific program of care and measure clinical and other outcomes"* [1]

Disease Management Programs (DMPs) are important, as pointed out by the Institute of Medicine because:

> *"The current delivery system responds primarily to acute and urgent health care problems....Those with chronic conditions are better served by a systematic approach that emphasizes self management, care planning with a multidisciplinary team and ongoing assessment and follow up"* [2]

DM is generally considered part of Population Health and is divided into the following categories:
- **Disease management**—focuses on specific diseases
- **Lifestyle management**—focuses on personal risk factors like smoking
- **Demand management**—focuses on improved utilization, as an example, emergency room usage
- **Condition management** —focuses on temporary conditions such as pregnancy and not diseases

Disease Management is discussed under Medical Informatics because it is dependent on information technology for several processes:

- Automated data collection and analysis
- Clinical Practice Guidelines (CPGs) that are web based or embedded into the electronic health record (EHR)
- Automated disease registries
- Telemonitoring of patients at home
- Patient tracking
- Networks to connect multiple healthcare workers on the DM team

Figure 10.1 illustrates how DMPs interact with other medical programs and information technology. In the future, CPGs, EBM programs and DMPs will be embedded or linked within all EHRs.

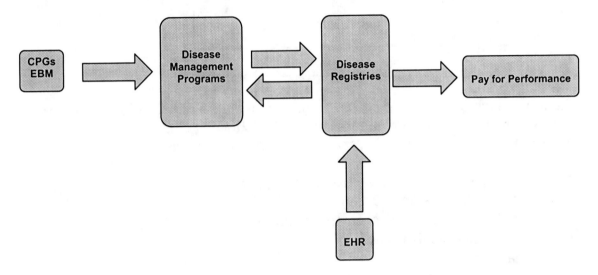

Figure 10.1. Inter-relationship between disease management programs, evidence based medicine (EBM), disease registries, pay for performance and electronic health records (EHRs)

DMPs evolved because health maintenance organizations (HMOs) wanted to control the rising cost of chronic diseases. The first DMP was established in the 1980's at Group Health of Puget Sound and Lovelace Health System in New Mexico and now are part of many large health care organizations. As an example, in a survey of over 1000 healthcare organizations, disease registries were established with the following frequencies: Diabetes (40.3%), Asthma (31.2 %), Heart Failure (34.8 %) and Depression (15.7 %). [3]

New attention may be paid to DMPs if pay for performance (P4P) becomes a reimbursement standard. Chronic diseases affect about 20% of the general population and account for 75% of health care spending. By the year 2030, 20% of the US population will be 65 or older. Chronic diseases are more likely to affect lower income populations who have limited access to medical care. Figure 10. 2 shows the predicted prevalence of chronic disease in the future. [4]

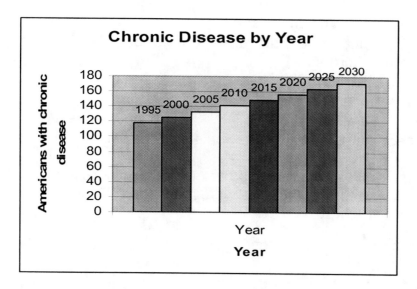

Figure 10.2. Predicted chronic disease prevalence (millions) by year

The most common chronic diseases to be managed are heart failure, diabetes and asthma due to high prevalence and cost. Following close behind are obesity, hypertension, chronic renal failure and obstructive lung disease.

Disease Management programs involve multiple players:
- Quality Improvement Organizations (QIOs)
- State and Federal Governments (Medicare and Medicaid)
- Pharmacy organizations
- Pharmaceutical companies
- Hospital Systems, including information technology
- Physicians and their office staff
- Employers
- Independent vendors; including EHR vendors
- Health Maintenance Organizations (HMOs)

The integration of multiple players is best demonstrated by the classic *Chronic Care Model* created by Dr. E. Wagner and the Macoll Institute for Healthcare Innovation. His model incorporates community resources, healthcare systems, information technology, patient participation and a disease management team.[5]

The usual processes involved in Disease Management are:
- Identification of a problem and a target population
- Comparison of local to national data
- Review of existing clinical practice guidelines to see if applicable CPGs can be used or modified
- Evaluation of patient self-management education
- Evaluation of process and outcomes measurement

- Feedback to clinicians and other hospital workers
- Emphasize systems, not individuals
- Coordination among multiple services and agencies

Prior to initiating a DM Program the following questions should be addressed:
- What is the goal? Decrease diabetic complications? Decrease trips to the emergency room by asthmatics?
- What population will be studied? Rural and urban? Insured and uninsured?
- Is the problem high volume or high cost or both?
- Are there preventable complications such as hospital admissions?
- How prevalent is the disease? Common enough to create a DMP?
- Are there practice variations among different medical groups?
- Are payers (insurance companies) interested?
- Do guidelines already exist or will a new one need to be created?
- Is the treatment feasible or practical?
- Are outcomes clearly defined and meaningful?
- Do Information systems already exist for the program? Data retrieval is easier if systems are already in place

The goal of all DMPs is to improve outcomes: clinical, behavioral, cost, patient functional status and quality of life.

Centers for Medicare and Medicaid's (CMS) position. As pointed out in other sections of this book, when CMS speaks, everyone listens because involvement may mean federal funding. A quote from the CMS web site:

> *"About 14 percent of Medicare beneficiaries have congestive heart failure but they account for 43 percent of Medicare spending. About 18 percent of Medicare beneficiaries have diabetes, yet they account for 32 percent of Medicare spending. By better managing and coordinating the care of these beneficiaries, the new Medicare initiative will help reduce health risks, improve quality of life, and provide savings to the program and the beneficiaries".*[6]

CMS has created 10 pilot programs to see if disease management can save the government money over a three year period (phase I). The Chronic Care Improvement Program (part of the Medicare Modernization Act of 2003) is now known as the Medical Health Support Program. Companies involved will not get paid for disease management unless they can show a total savings of 5% compared to a control group. Companies that can demonstrate improved outcomes are asked to participate in phase II and will likely tackle diabetes or heart failure. The companies selected were: American Healthways, XL Health, Health Dialog Services, LifeMasters and McKesson Health Solutions. All will need robust information technology to succeed.[6]

What do we know from current programs and the medical literature?

- A study in the Journal of the American Medical Association (JAMA) demonstrated that 32 of 39 interventions showed improvement in at least one process or outcome measurement for diabetic patients; 18 of 27 studies involving three chronic conditions also demonstrated lower health care costs and/or lower utilization of services [7]

- A comprehensive DMP for African-American diabetics showed large reductions in amputations, hospitalizations, emergency room visits and missed work days [8]

- HealthPartners Optimal Diabetes Care Impact: 400 fewer cases of retinopathy (eye damage) each year; 120 fewer amputations each year; 40-80 fewer myocardial infarctions (heart attacks) per year [9]

- A systematic review/meta-analysis of DMPs on heart failure concluded: programs are effective in reducing admissions in elderly patients [10]

- A DMP program for myocardial infarctions reduced readmissions, emergency room visits and insurance claims [11]

- MaineHealth. By using disease registries and focused care, Maine's largest hospital system was able to realize more than $1 million savings annually with reduced emergency room visits and hospitalizations. They focused on asthma, diabetes, depression and heart failure. Physician teams used national asthma guidelines and receptionists used registries to call and remind asthmatics about appointments and flu shots. Visits to emergency rooms dropped from 55% to 16% in certain disease categories [12]

Disease Registries

Definition of a disease registry:

> "A software application for capturing, managing and providing access to condition specific information for a list of patients to support organized clinical care" [13]

Registries are tools that disease management programs use to track patients with chronic diseases such as diabetes. As a result, they can remind patients to get lab work done and keep appointments. In addition, they can aggregate data to show, for example, average hemoglobin A1c levels (blood test to measure blood sugar control) that could be useful for "pay for performance" programs.

Disease registries are available in several formats:
- **Manual:** data manually inputted into paper or a computer database or spreadsheet or into a web based program
- **Automatic:** data automatically inputted into standalone software or web based site using client-server software and integrated with, for example, a laboratory result program
- **Automated and Integrated:** data input, retrieval, tracking and graphing are all automatic and part of an electronic health record

Potential drawbacks of registries include manual inputting, need for accurate coding, need for frequent updating and need for additional staff to maintain a registry. Disease management team members need to meet on a regular basis to discuss data reports and analysis. [14]

Approximately 50 disease registries exist that are free or fee based. Cost is usually $500-$600 year per user for commercial registries. In general, free public registries have less functionality than commercial registries. For an excellent in-depth review of 16 registries see Chronic Disease Registries: A Product Review. [15] Another excellent review is available through the Virginia Health Quality Center.[16] For an excellent review of IT tools for chronic disease management, we recommend the July 2006 monograph by the California HealthCare Foundation. [17]

Disease Registry Examples

Chronic Disease Electronic Management Systems (CDEMS). This popular program is Microsoft Access based and tracks diabetes and adult preventive health. The program is customizable and includes lab reminders for clinicians. The reports generated are also customizable and users have access to a web forum to discuss issues. A free add on program inputs data automatically from several laboratory information systems (Quest, Labcorp, Dynacare and PAML) Shortcomings include the need to manually input data and access is limited to 10 concurrent users.[18]

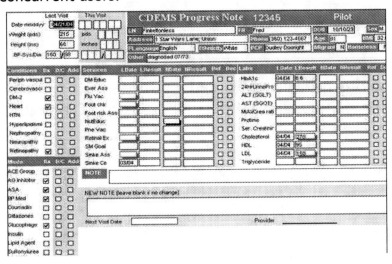

Figure 10.3. CDEMS disease registry (courtesy Washington State Department of Health - Diabetes Prevention & Control Program and Centers for Disease Control - Diabetes Translation Division)

Population Health Navigator (PHN). Population Health Navigator is a program used by the Department of Defense to track asthma, beta-blocker use following myocardial infarction, cardiovascular risk factors, breast cancer screening, cervical cancer screening, depression, diabetes, hypertension, COPD, hyperlipidemia, low back pain and high utilizers. Data can be analyzed by physician, clinic or hospital system. Drawbacks include that it is not integrated into the electronic health record and data is not available real time. The site is secure and only available to DOD personnel with proper authority.[19]

DocSite PatientPlanner. One of the best known web based commercial registries is Patient Planner by DocSite that will track multiple common diseases. It can be integrated with practice management software, EHRs and e-prescribing systems. Below is a typical clinician report with lab results and due dates. Clinical practice guidelines can be embedded in the registry with the ability to make local modifications. Other features include HL7 links to input lab data, patient education, patient letter generation and the ability to host data locally or on the DocSite server.[20]

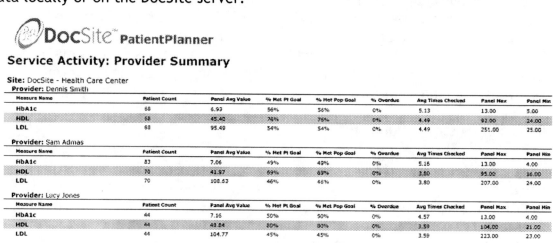

Figure 10.4. DocSite PatientPlanner registry (courtesy DocSite)

EClinicalWorks. True disease registries are uncommon as part of current electronic health records. This is the ideal format in that patient lists, alerts, reminders, patient education, lab results, CPGs, patient education and reporting could all be part of one information system. The screen shot below is from the registry feature of the eClincalWorks EHR. [21]

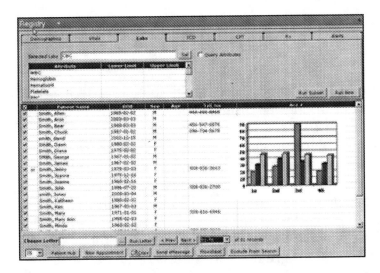

Figure 10.5. EClinicalWorks registry (courtesy EClinicalWorks)

CareManager. CareManager by Kryptiq integrates a disease management program into existing EHR systems. Data therefore approaches real time as opposed to programs based on insurance claims data. Currently, they offer modules for diabetes, stroke prevention and coronary heart disease. In development are cancer screening, tobacco cessation and osteoporosis. As demonstrated in figure 10.6 patient status is color coded (green = good, etc). A pilot project with Providence Medical Group demonstrated a ROI in only 4.5 months as a result of more outpatient visits and a higher level of coding. They also demonstrated improved compliance with multiple diabetic measures such as cholesterol and blood pressure control. Disease dashboards demonstrating their status can be e-mailed to patients. Organizations and physicians can be given performance score cards to mark individual progress. Managers can customize reports for pay for performance programs. [22]

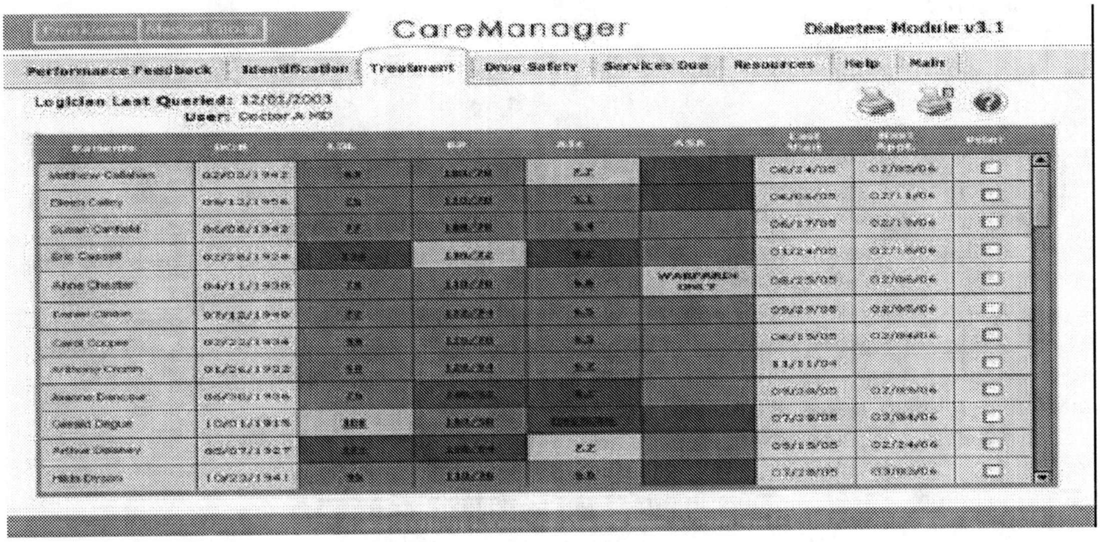

Figure 10.6. CareManager registry (courtesy Kryptiq)

Conclusion

For Disease Management programs to succeed there needs to be a general mandate to improve the treatment of chronic disease and financial support. Due to the rising costs for chronic diseases, CMS and managed care organizations are interested in new pilot programs. What must be shown is that DM programs improve patient outcomes and save money. The Congressional Budget Office in Oct 2004 concluded that there is inadequate evidence that DM programs reduce healthcare spending.[23] In some instances they might actually increase costs if more services are recommended. Proof of long term benefit is evolving at this time. Nevertheless, we believe that information technology can aid the study of diseases and populations greatly. Ultimately, all electronic health records will have comprehensive disease management features that will be customizable for clinicians and administrators. Data will be easy to retrieve in a real time mode and will be linked to reimbursement. Until that happens, however, we will rely on a variety of disease registries and disease management systems

References

1. Epstein RS, Sherwood LM. 1996. From outcomes research to disease management: a guide for the perplexed. Ann Intern Med 124: 832-837
2. Crossing the Quality Chasm: A new health system for the 21th century. 2001. National Academies Press http://www.nap.edu/books/0309072808/html (Accessed March 5 2006)
3. Casolino L, Gillies RR, Shortell SM, *et al*. External incentives, information technology, and organized processes to improve health care quality for patients with chronic diseases. *JAMA*. 2003;289: 434-41.
4. Wu, Shin-Yi and Green, Anthony. *Projection of Chronic Illness Prevalence and Cost Inflation*. RAND Corporation, October 2000
5. Chronic Care Model http://www.improvingchroniccare.org/change/model/components.html (Accessed March 2 2006)
6. Centers for Medicare and Medicaid Services http://www3.cms.hhs.gov/apps/media/press/release.asp?Counter=1274 (Accessed March 2 2006)
7. Bodenheimer T, Wagner, E H, Grumbach K Improving Primary Care for Patients With Chronic Illness: The Chronic Care Model, Part 2 JAMA 2002;288:1909-1914
8. Patout CA et al Effectiveness of a comprehensive diabetes lower extremity amputation prevention program in a predominately low income African-American population Diabetes Care 2000;23:1339-134
9. HealthPartners. Dr Gail Amundsen (personal communication)
10. Gonseth J et al The effectiveness of disease management programmes in reducing hospital admissions in older patients with heart failure: a systematic review and meta-analysis of published reports Eur Heart Journal 2004;25:150-95
11. Young W et al A disease management program reduced hospital readmission days after myocardial infarction CMAJ 2003;169:905-10
12. Report: Chronic care treatment improves patient's health, lowers cost www.ihealthbeat.org November 10 2004 (Accessed March 2 2006)

13. Using Computerized Registries in Chronic Disease
http://stage.chcf.org/documents/chronicdisease/ComputerizedRegistriesInChronicDisease.pdf
(Accessed March 5 2006)
14. Using Computerized Registries in Chronic Disease
http://stage.chcf.org/documents/chronicdisease/ComputerizedRegistriesInChronicDisease.pdf
(Accessed March 5 2006)
15. Chronic Disease Registries: A Product Review May 2004 www.chcf.org (Accessed March 5 2006)
16. Registry Products http://www.vhqc.org/inc/pdf/regprodgrid.doc (Accessed March 10 2006)
17. Jantos LD, Holmes ML IT Tools for Chronic Disease Management: How do they measure up? California HealthCare Foundation July 2006 www.chcf.org (Accessed January 26 2007)
18. Chronic Disease Electronic Management Systems www.cdems.com (Accessed March 6 2006)
19. Navy Environmental Health Center. Navy Population Health Navigator. http://www-nehc.med.navy.mil/hp/PH_Navigator/FAQ.htm#PHN1 (Accessed March 19 2006)
20. DocSite http://www.docsite.com/ (Accessed March 5 2006)
21. eClinicalWorks http://www.eclinicalworks.com/index.php (Accessed March 10 2006)
22. Kryptiq http://www.kryptiq.com/ProviderSolutions/DiseaseManagement.html (Accessed March 27 2006)
23. Congressional Budget Office http://www.cbo.gov/showdoc.cfm?index=5909&sequence=0 (Accessed March 5 2006)

Chapter 11: Pay for Performance (P4P)

Learning Objectives

After reading this chapter the reader should be able to:
- State the origins behind pay for performance programs
- Describe the need for automated retrieval of accurate patient data for the success of pay for performance programs
- List governmental pay for performance pilot programs
- List the concerns and limitations of current pay for performance programs for the average clinician

Introduction

There have been numerous studies since the classic *Crossing the Quality Chasm* that confirm we are not getting our money's worth from American Medicine. Furthermore, a study by the Commonwealth Fund, demonstrated that the quality of care delivered to Medicare recipients was not related to the amount of money spent.[1] The Institute of Medicine (IOM) has been consistently critical of the variance in care and serious patient safety issues. As a result, they have repeatedly called for an increase in payments to clinicians who offer higher quality care. These concerns about value based care are further aggravated by the fact that the United States has a $1.9 trillion dollar health care price tag that continues to rise each year. Most recently the IOM released "Rewarding Provider Performance: Aligning Incentives in Medicare" in September 2006 that had the following key messages:
- "Fundamental change requires a commitment by all Medicare providers to deliver high quality care efficiently
- Pay for performance constitutes one key component needed for the transformation of the healthcare payment system but cannot achieve this transformation alone
- Pay for Performance offers significant promise and can begin now by building off other strategies for improvement
- In particular, providers should assume accountability for transitions between settings of care and coordinate care in treating patients with chronic diseases
- Pay for performance in Medicare should be introduced within a learning system that has the capacity to assess early experiences, adjust for unintended consequences and evaluate impact" [2]

All of these factors have helped give birth to the notion that we need major changes in the field of medicine, to include how we determine reimbursement for care. No incentives for better quality exist under our current system. The more widgets you make, the more you get paid. The widgets don't have to be made well, just well documented. Using information technology data from electronic claims, electronic health records and disease registries we can measure quality parameters faster and without the need to do paper chart reviews. In fact, some authorities feel that P4P should first pay to create an information technology infrastructure and later reimburse for quality.

P4P (also known as value based purchasing) has gained traction in the United States in a surprisingly short period of time. The momentum may in part be due to the 2004 statement made by Mark McClellan, administrator for the Center for Medicare and Medicaid Services in the Wall Street Journal:

> "In the next five to ten years, pay for performance based compensation could account for 20-30 percent of what the federal programs pay providers" [3]

A 2006 article in the New England of Medicine examined the incidence of P4P programs in 252 Health Maintenance Organizations (HMOs). They determined that over half had P4P programs; 90% of programs were for physicians and 38% were for hospitals. [4]

There have been several bills introduced in Congress to address pay for performance but only one has passed thus far. The Tax Relief and Health Care Act of 2006 (HR 6111) implemented a voluntary quality reporting system for Medicare payments tied to claims data. Clinicians who report this information would be eligible for a 1.5% bonus. This program will begin in July 2007 and end in December 2007. New measures are likely for 2008 and data can be submitted from disease registries. The new system is called the Physician Quality Reporting Initiative. [5]

The concept of P4P is not entirely new as WellPoint (a health plan covering 24 million lives) has had reimbursement based on clinical measurements for about 10 years, but it was not called P4P. [6] Additionally, HealthPartners (Minnesota) has had an Outcomes Recognition Program since 1997. [7]

The principles behind Pay for Performance are aimed at better:
- Quality of patient care
- Patient and clinician satisfaction
- Patient safety
- Reimbursement to clinicians
- Return on investment (ROI)

Examples of data that might help evaluate physician performance:

Types of data	Examples
Utilization data	Emergency room visits
Clinical quality	Women who have had mammograms
Patient satisfaction	Percent of patients who would recommend their primary care manager
Patient safety	Percent of patients questioned about allergic reactions

Table 11.1. Types and examples of data for P4P programs

Information Technology Issues:

- Most electronic health records (EHRs) are not ready for generating P4P type reports. Ideally, data would be automatically generated from the EHR if the data was inputted into data fields in templates rather than free text. Unfortunately most notes are not written using an electronic template. Also, data can't come from problem summary lists as they are not updated often enough. Perhaps artificial intelligence will eventually be able to scan a dictated patient encounter and automatically submit a P4P report as well as a coding level
- A February 2007 article on automated review of quality measures for heart failure using a EHR, concluded that the current system lacked the ability to tell why a drug was not started or why is was stopped. Chart reviews were the only way to tell why recommended medications were not used or discontinued [8]
- There is a need to identify acute versus chronic problems and active versus inactive problems
- Until EHRs are universal, organizations must have a transitional plan like disease registries and disease flow sheets
- Many healthcare systems would benefit from a central data repository (CDR) with a rules engine. [9] Data could be pushed or pulled from the CDR for monthly reports

The Ambulatory Care Quality Alliance (ACQA). In order for P4P to be well received there needed to be a set of outpatient clinical performance measures that would be accepted by clinicians. The Ambulatory Care Quality Alliance, the American Academy of Family Physicians, The American College of Physicians, America's Health Insurance Plans and the Agency for Healthcare Research and Quality met in January 2005 and recommended 26 "starter set" measures.[10] The program will look at *processes* and not actual *patient outcomes* at this time. This will be similar to the Health Plan Employer Data and Information Set (HEDIS) measurements that are commonly used today as performance measurements.[11] As a rule, process measurements will check to see if a test was done and not the actual result. This should allow for easier retrieval of data using administrative or insurance claims data. The goal is to eventually have national ambulatory quality measures in place.
The following are the categories of the proposed starter set measures:

- Preventive measures

- o Example: Percent of women who had mammograms or Pap smears
- Coronary heart disease
- Heart failure
- Diabetes
- Asthma
- Depression
- Prenatal care
- Measures to address overuse or misuse [10]

Pay for Performance Projects

It is estimated that more than 80 organizations will have P4P in place by 2006 in spite of the paucity of studies to prove efficacy or return on investment.

Centers for Medicare and Medicaid Services (CMS) Premier Hospital Quality Incentive Demonstration Project

- 270 hospitals have been participating since October 2003
- Hospitals are paid based on compliance with 34 quality indicators in 5 common areas (heart attack, heart failure, pneumonia, bypass surgery and hip/knee disease)
- $7 million given out yearly for 3 years as bonuses
- Three year demonstration project
- Program uses the Premier Perspective database, the largest in the nation, currently with 3 billion patient charge records
- After review of first year data, hospitals scoring in the top 10% will be given a 2% bonus in Medicare payments. Hospitals scoring in the second 10% will get 1% and those below will get nothing
- It is possible for hospitals to have a 1-2% decrease in Medicare payments if at year three they have not improved beyond the baseline
- $8.85 million awarded to top 123 performers
- Demonstration project for nursing homes is likely in the future
- Actual improvements in quality measures documented in the first year of the project:
    - o Increase from 90 percent to 93 percent for patients with acute myocardial infarction (heart attack)
    - o Increase from 86 percent to 90 percent for patients with coronary artery bypass graft
    - o Increase from 64 percent to 76 percent for patients with heart failure
    - o Increase from 85 percent to 91 percent for patients with hip and knee replacement
    - o Increase from 70 percent to 80 percent for patients with pneumonia [12]

- A two year report was published in February 2007 in the New England Journal of Medicine and showed:
  - After adjusting for baseline performance, P4P programs were associated with improvements in the 2.6-4.1% range, compared to hospitals that reported but were not part of the program
  - It is unknown what the actual return on investment was for the average participating hospital
  - Patient outcome information is unknown. In other words, did better compliance to guidelines result in fewer deaths and complications?
  - Would the results have been better with a higher incentive? [13]

Centers for Medicare and Medicaid Services (CMS) Medicare Care Management Performance Demonstration

- Passed in late 2006 as part of section 649 of the 2003 MMA
- Three year demonstration project for small to medium sized practices in the states of Arkansas, California, Massachusetts and Utah
- In the first year practices will only report baseline quality data and will be paid for reporting
- Reimbursement could be up to $10,000 per physician and $50,000 per practice
- Implies they need information technology for data retrieval and reporting
- Reimbursement levels may be too low [14]

Surgical Care Improvement Project

- Similar Medicare P4P project for surgical care
- Will look at post-surgery: site infections, adverse cardiac events, deep vein thromboses (blood clots in the legs) and pneumonia
- Program began in July 2005 [15]

Bridges to Excellence

- Is a consortium of employers (General Electric, Procter and Gamble, Verizon Communications, Raytheon Company, UPS, Humana, Ford Motor Company and Cincinnati Children's Hospital Medical Center), health plans (Aetna, Anthem Blue Cross Blue Shield of Ohio and Kentucky, Blue Cross Blue Shield of Illinois, Alabama and Massachusetts, Tufts Health Plan, United Healthcare, Harvard Pilgrim Healthcare and Humana) and physician groups in 10 states
- Their goal is to raise quality and drive down costs
- Program paid out about $2 million in 2005 to clinician groups who adopt P4P [16]

Providence Health Systems
- Has 17 hospitals in California, Oregon, Washington and Alaska
- Since 2003, PHS has collaborated with Kryptiq (data integrating company) to develop a tool that uses EHR data for disease management
- Plans to have about 15 disease modules
- Improvement in compliance to national guidelines already seen [17]

CareFirst Blue Cross/Blue Shield
- Will pay up to $20 thousand for installing EHRs
- Part of Bridges to Excellence program
- Plans to spend about $3.6 million over next 3 years for P4P [18]

Blue Cross/Blue Shield of Michigan
- Paid $1 million in bonuses in April 2005 to physicians who encouraged patients to use less expensive drugs and follow clinical practice guidelines (CPGs)
- Approximately 2,400 physicians enrolled
- On a trial basis physicians were given 0.5% less in 2006 to create a pool of money [19]

California Pay for Performance Collaboration
- Seven major insurance companies and 225 medical groups have organized towards P4P since 2000
- P4P bonuses totaled $100 million in 2004
- Program uses aggregate insurance claims data that is publicly reported and managed by an independent source
- Funded through multiple sources
- Second year data showed gains in all areas of clinical quality
- Groups with more IT did better
- The 2005 measures for P4P are divided into the following categories:
    - Prevention and Disease Management 50%
    - Patient Satisfaction 30%
    - Information Technology 20% [20]

HealthPartners Outcomes Recognition Program
- Began in 1997 and later called P4P in 2001
- In 2004 it paid $5.6 million to primary care groups, $1 million to specialty groups and $4 million to hospitals
- Payouts: 25% for patient satisfaction, 75% for measures of diabetes, coronary disease, preventive services, tobacco cessation and use of generic drugs
- Pays for outcome measures and not just process improvement [21]

The British Experience

- Although some P4P programs began in 1990, the current extensive program has existed only since April 2005
- Deals with 10 chronic conditions
- Most data will come from practitioner's computers and the PRIMIS (primary care information services) that provides free of charge:
    - **Training** in information management skills and recording for data quality
    - **Analysis** of data quality, plus a comparative analysis service focused on key clinical topics
    - **Feedback** and interpretation of the results of data quality and comparative analyses
    - **Support** in developing action plans [22]
- Practitioners operate on a points system with a possible total of 1050 points gained
- Program could result in potentially as much as $77,000 per physician [23]

Clear Choice Health Plans

- Opted to use three measures of quality for P4P awards program:
    - Use of evidence based medicine by accessing the medical resource UpToDate
    - Appropriate ordering of images by accessing the American College of Radiology web site that lists appropriateness criteria
    - Self improvement by accessing a reporting web site www.managedcare.com [24]

P4P Concerns and Limitations

The following are some of the concerns about P4P programs expressed primarily by physicians and their organizations.

- Does it discriminate against practices without EHRs?
- Are EHRs sophisticated enough to provide accurate measures of quality?
- Should data be public?
- Should reporting be voluntary?
- Will it cause clinicians to "dump" non-compliant patients?
- Will it result in higher quality care or long term return on investment?
- Will it adjust for sicker, poorer, more elderly patients?
- Much of the practice of medicine does not have identified quality measures, so P4P will not apply
- Will the motive be money and not quality?
- Is a bonus of 5-10% of yearly compensation adequate for P4P programs?
- Is the extra work to report actually worth the small payment?
- Do P4P programs favor large medical groups?
- Should data come from a centralized data repository?

- Will P4P work outside health maintenance organizations (HMO's)?
- Should bonuses be paid for improvement even if results do not meet national goals?
- At this time, the majority of P4P reimbursements go to primary care physicians and not specialists or hospitals
- Waiting on "report cards" occasionally takes a long time and impedes next year's improvement [25-31]
- Physician's skepticism as depicted in figure 11.1

Medscape Physician Survey [32]

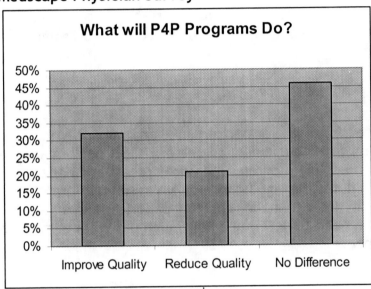

Figure 11.1. Medscape Survey on P4P

Current status of pay for performance programs

On a positive note, P4P has the potential to blend evidence based medicine, disease management, clinical practice guidelines, electronic health records and information technology. Nevertheless, P4P programs have their share of obstacles. There have been multiple P4P articles written that are both pro and con in the lay press, but little written in the medical literature. In a 2005 article by Rosenthal et al in the Journal of The American Medical Association, they reported a study comparing P4P in a large California physician group compared to a control group. The measurements studied were Pap smears, mammograms and Hgb A1C (diabetic) testing. There was very little improvement noted except for modest improvement in Pap smear testing. In spite of the $3.4 million payout by the health plan, the conclusion of the study was *"Paying clinicians to reach a common, fixed performance target may produce little gain in quality for the money spent and will largely reward those with higher performance at baseline".* [33] A 2006 systematic review by Petersen et al, was unable to prove that P4P incentives have been shown to improve the quality of care. [34] Until better evidence is available we should proceed cautiously and be sure performance measures are fair and equitable as outlined by the Medical Group Management Association (MGMA) 2005 principles:

- The primary goal must be improving health quality and safety
- Participation by practices should be voluntary
- Physicians and professional organizations should be involved in P4P design
- P4P measures must be evidence based, broadly accepted, clinically relevant and continually updated
- Physicians should have the ability to review and correct performance data
- P4P must reimburse physicians for any administrative burden
- P4P must reward physician's use of electronic health records and decision support tools [35]

Conclusion

The jury is out when it comes to pay for performance programs. No one disputes the need to change reimbursement to better reflect performance and not just volume. Current fledgling programs measure process and not actual patient outcomes which dilutes the significance of any results. Information technology is mandatory in order to make reporting paperless and effortless. Also, further research will need to determine how much pay will be required for how much performance.

References

1. Leatherman S, McCarthy, D. Quality of Health Care for Medicare Beneficiaries: A Chartbook 2005. The Commonwealth Fund http://www.cmwf.org/publications/publications_show.htm?doc_id=275195 (Accessed March 2 2006)
2. Rewarding Provider Performance: Aligning Incentives in Medicare IOM September 2006 www.iom.edu (Accessed October 22 2006)
3. Wall Street Journal September 17 2004 (Accessed October 15 2005)
4. Rosenthal MB et al. Pay for performance in commercial HMOs. NEJM 2006;355:1895-902
5. CMS Quality/Pay for Performance Initiatives. https://www.do-online.org/index.cfm?PageID=gov_regqualityperform (Accessed February 19 2007)
6. WellPoint www.wellpoint.com (Accessed November 2 2005)
7. HealthPartners www.healthpartners.com (Accessed November 2 2005)
8. Baker, DW, Persell SD, Thompson JA et al. Automated Review of Electronic Health Records to Assess Quality of Care for Outpatients with Heart Failure. Ann Intern Med 2007;146:270-7
9. White paper: Pay for performance Information Technology Implications for Providers. First Consulting Group Feb 2005 www.fcg.com (Accessed November 20 2005)
10. Agency for Healthcare Research and Quality. Recommended Starter Set. http://www.ahrq.gov/qual/aqastart.htm (Accessed November 7 2005)
11. *2003 National Study of Provider Pay-for-Performance Programs: Lessons Learned.* San Francisco: Med-Vantage Inc.

12. CMS/Premier Hospital Quality Incentive Demonstration Project
http://www.cms.hhs.gov/HospitalQualityInits/downloads/HospitalPremierFS200602.pdf
(Accessed January 2 2006)

13. Lindenauer PK et al. Public Reporting and Pay for Performance in Hospital Quality Improvement. NEJM 2007;356:486-496

14. CMS Demonstration Site
http://www.cms.hhs.gov/DemoProjectsEvalRpts/MD/itemdetail.asp?filterType=keyword&filterValue=medicare%20care&filterByDID=0&sortByDID=3&sortOrder=ascending&itemID=CMS057286
(Accessed December 24 2006)

15. Martin CB Medicare's Pay for Performance Legislation: A newsmaker interview with Thomas Russell MD www.Medscape.com   September 15 2005 (Accessed February 20 2006)

16. Bridges to Excellence
www.bridgestoexcellence.org/bte/docs/CapitalRegionBTEPressRelease_Final.doc (Accessed February 17 2006)

17. Endrado P. Pay for performance tools evolve as market shifts www.healthcareitnews.com September 26 2005 (Accessed February 9 2006)

18. Pay for performance www.acponline.org/weekly/2005/4/5/index.html (Accessed February 22 2006)

19. Merx K Win-win program for docs, patients. Detroit free press. April 25 2005 www.freep.com (Accessed February 1 2006)

20. California Pay for Performance Collaboration
http://www.pbgh.org/programs/documents/PBGH_ProjSummary_P4P_03_2005.pdf (Accessed February 11 2006)

21. Apland BA, Amundson GM. Financial Incentives, an indispensable element for quality improvement. Patient Safety & Quality Healthcare. Sept/Oct 2005. www.psqh.com (Accessed February 2 2006)

22. Primus + University of Nottingham http://www.primis.nhs.uk/pages/default.asp (Accessed November 20 2006)

23. Roland M Linking Physicians' Pay to the Quality of Care — A Major Experiment in the United Kingdom NEJM 2004;351:1448-1454 (Accessed February 7 2006)

24. Patmas MA. A Novel Pay for Performance Program www.uptodate.com/p4p.ppt (Accessed February 20 2006)

25. Rohack JJ The Role of Confounding Factors in Physician Pay for Performance Programs. Johns Hopkins Advanced Studies in Medicine 2005;5:174-75 (Accessed January 20 2006)

26. Audet AM et al Measure, Learn and Improve: Physicians' Involvement in Quality Improvement Health Affairs 2005;24:843-53

27. Raths D Pay for Performance. Healthcare Informatics Feb. 2006; 48-50

28. Colwell J Market forces push pay for performance. ACP Observer May 2005

29. Shaw G What can go wrong with pay for performance incentives ACP Observer March 2006

30. Colwell J Pay for performance takes off in California ACP Observer Jan/Feb 2005

31. Bodenheimer T et al Can money buy Quality? Physician response to pay for performance.
http://hschange.org/CONTENT/807/ (Accessed February 23 2006)

32. Medscape Physician Poll on P4P March 2006 www.medscape.com (Accessed March 20 2006)

33. Rosenthal MB et al .Early Experience With Pay-for-Performance: From Concept to Practice JAMA. 2005;294:1788-1793

33. Petersen LA et al. Does Pay for Performance Improve the Quality of Health Care? Ann of Int Med 2006;145:265-272

35. MGMA. http://mgma.com/press/MGMApospayforperformance.cfm (Accessed February 4 2006)

Chapter 12: Patient Safety

Learning Objectives

After reading this chapter the reader should be able to:
- Identify why patient safety is a national concern
- List the perceived causes of patient safety concerns
- Describe the role of information technology in improving patient safety
- Compare and contrast the private and governmental patient safety programs
- List the various technologies that are likely to improve medication error rates such as computerized physician order entry
- Identify the obstacles to widespread implementation of patient safety initiatives

Introduction

The following statement was posted by a bioengineer after the tragic death of Betsy Lehman, a well known Boston columnist who died from an excessive dose of chemotherapy in 1995:

> *"How long, Oh Lord, must this continue? In 1974 we had an on-line patient record system that flagged unusual lab results or unusual...prescriptions, and that was at a vet[erans] hospital. That's 21 years ago...Isn't it time that basic computerization be part of the expected, and required, care at medical facilities? That humans make 0.1 percent errors on prescriptions may be forgivable; that hospitals don't take obvious actions to protect themselves and patients, well within state-of-the-art, is not"* [1]

Another sobering statistic about United States healthcare is that those activities that result in more than one death per 1000 encounters include bungee jumping, mountain climbing and healthcare. Dr Lucian Leape from the Harvard School of Public Health in 1994 estimated that 180,000 patients die each year as a result of medical errors which is the equivalent of three jumbo jets crashing every two days. Surprisingly, he also pointed out that a literature search in 1992 resulted in no articles on preventing medical errors. According to Dr. Leape, the only specialty in Medicine that has experienced dramatic advances in patient safety is Anesthesiology with less than one death in 200,000 patients undergoing anesthesia. [2] Other industries such as the airlines have dramatically reduced mishaps thru initiatives such as "crew resource management". As an example, Southwest Airlines has had no mishaps in over 9.5 million flights. [3] Besides the obvious increased mortality and morbidity that results from medical errors there is a resulting increase in litigation. It was estimated in 2003 that malpractice costs totaled $27 billion. [4]

One of the basic premises of this chapter is:
- Technology can improve the quality of medical care
- Improved quality of care means improved patient safety
- Technology can therefore improve patient safety

What is missing, unfortunately, is high quality research that proves this premise. Let us start to look at patient safety further by highlighting some of the shortcomings of medicine currently practiced in the United States today.

The United States Medical Report Card

World Health Organization (WHO). Table 12.1 shows that the US spends a greater percent of the gross national product (GNP) on healthcare compared to other countries but life expectancy is not better as depicted in table 12.2.

Country	1997	1998	1999	2000	2001
United Arab Emirates	3.6	4	3.7	3.5	3.5
United Kingdom	6.8	6.9	7.2	7.3	7.6
Tanzania	4.1	4.4	4.3	4.4	4.4
USA	13	13	13	13.1	13.9
Uruguay	10	10.2	10.7	10.9	10.9

Country	Age
United Arab Emirates	72.5
United Kingdom	78.2
Tanzania	46.5
USA	77.3
Uruguay	75.2

Table 12.1. Percent of Gross National Product spent on healthcare

Table 12.2 Life expectancy in years

Based on available outcome measures, the United States is generally in the bottom half and its relative ranking has been declining since 1960. [5]

Rand Study 2003
Some have argued that one of the causes of decreased patient safety is the failure of US physicians to follow national clinical practice guidelines. A telephone survey of 13,275 adults living in 12 urban areas in the US was conducted that looked at quality indicators for 30 acute and chronic conditions including preventive care. The conclusion was that participants received only 55% of recommended care. It is unknown, however, how often patient non-compliance or lack of finances played a role in patients not receiving the recommended care. [6]

Institute of Medicine (IOM) Reports
The 1999 Institute of Medicine report *To Error is Human* estimated that at least 98,000 inpatients die every year and 1,000,000 are injured due to preventable errors. [7] The mortality and morbidity rate may have been actually higher as

outpatient adverse events were not reported. [8] The 2001 Institute of Medicine report *Crossing the Quality Chasm* stated that there has been little progress since the first report. No congressional action has taken place since the first IOM report. The current medical system was described as an "era of Brownian motion in health care". Furthermore, the IOM commented on three categories of medical errors:

- *Overuse:* examples are widespread use of antibiotics for viral infections; 32% of carotid artery surgeries were clearly inappropriate and 32% were equivocal
- *Under use:* examples are vaccine use, cancer screening, beta blockers post heart attack, etc
- *Misuse:* examples are 36% of patients with active tuberculosis were not treated with four drugs initially, as recommended

The IOM has long been an advocate of using information technology to improve healthcare, particularly patient safety. The 2001 IOM Executive Summary recommended that we " improve access to clinical information and support clinical decision making" and "create national information infrastructure to improve health care delivery and research". Also, their goal is to eliminate handwritten notes in the next decade. [9] The IOM's 2004 *Patient Safety: Achieving a new standard for Care* repeated the same recommendations. [10]

HealthGrades 2007 Hospital Quality in America Study

This organization reviewed 41 million Medicare patient records from 5,000 hospitals from 2003-2005. They concluded that although mortality continues to improve there were wide gaps between the best and the worst hospitals. There was a 69% lower rate of dying in the best compared to the worse hospitals studied. The rate of hospital acquired infections correlated best with overall performance. The leading 6 safety and quality issues measured were:

- Post operative sepsis
- Post-operative respiratory failure
- Decubitus ulcers (bed sores)
- Post-operative pulmonary emboli or DVTs (blood clots)
- Hospital acquired infections
- Failure to rescue (not recognizing and treating a deteriorating patient)

Samantha Collier MD of HealthGrades believes that the hospitals that traditionally have excellent safety scores have a "culture of safety" and they are the ones that have all of the mechanisms including technology in place to prevent and track patient safety issues [11]

Commonwealth Fund Study 2005

This was a Survey of more than 700 adults from the United States, Australia, Canada, Germany, New Zealand and the United Kingdom. The US reported the highest number of medical errors and 50% of US adults stated they went without medical care in past year due to high cost; as compared to 13% for the United Kingdom. [12]

Why is the USA Healthcare report card unfavorable?

In Crossing the Quality Chasm the IOM states the problem of poor medical care is based on the:
- Growing complexity of science and technology
- Increase in chronic conditions, e.g. obesity, diabetes and heart failure
- Poorly organized delivery systems that are not organized around patient safety
- Constraints on exploiting the revolution in information technology [13]

A well known healthcare consultant Dr William Yasnoff states that medical care in the United States is sub-standard because:
- Medical error rates are too high
- Healthcare quality is inconsistent
- Medical research results are not rapidly used
- Healthcare costs are escalating
- New technologies continue to drive up costs
- Baby boomers will greatly increase demand
- Capacity for early detection of bioterrorism is minimal [14]

As pointed out by the Agency for Healthcare Research and Quality, there is too much variation in medical care within the United States as demonstrated by variation in coronary angiography rates vary between states (Fig. 12.1). [15]

Coronary Angiography

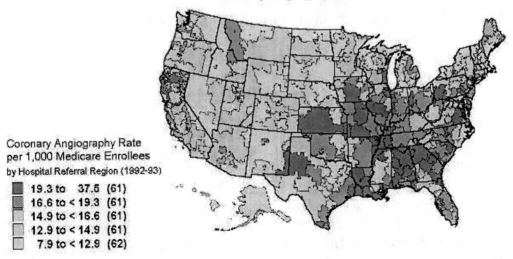

Coronary Angiography Rate
per 1,000 Medicare Enrollees
by Hospital Referral Region (1992-93)

- 19.3 to 37.5 (61)
- 16.6 to < 19.3 (61)
- 14.9 to < 16.6 (61)
- 12.9 to < 14.9 (61)
- 7.9 to < 12.9 (62)

Figure 12.1. US Coronary Angiography Rate (adapted from AHRQ web site)

In addition there is a significant difference between early landmark trials producing national recommendations and widespread implementation as shown in Table 12.3.[16]

Clinical Procedure	Landmark Trial	Current rate of use
Flu vaccine	1968	64% (2000)
Pneumococcal Vaccine	1977	53% (2000)
Diabetic eye exam	1981	48.1% (2000)
Mammography	1982	75.5% (2001)
Cholesterol Screening	1984	69.1% (1999)

Table 12.3. Current Rate of Use of Landmark Trial Results

Barriers to Improving Patient Safety through Technology

- **Organizational.** US Medicine is primarily a decentralized system with no unifying philosophy. Many small physician groups have no loyalty to hospitals or other healthcare organizations

- **Financial.** Who will pay for what? It is estimated that it will cost $500-700 billion dollars over the next 10 years to have a full fledged interoperable electronic health record nation wide. This is 3-4% of the total health care budget which is a lower percentage than what other industries spend on technology. In 1996 the healthcare industry spent about $543 per worker as compared to $12,666 per worker spent by security brokers and other industries for information technology [17]

- **Technical.** How can we train and implement technologies such as electronic health records for everyone?

- **Privacy.** HIPAA concerns continue to be an obstacle to many developments in technology

- **Security.** Several major episodes of identify theft and the loss of medical records will likely slow the process of adopting electronic health records and other technologies

- **Behavioral.** Some would argue that the medical profession has "mural dyslexia" or failure to see the handwriting on the wall. This is due to a fear of change and an underlying skepticism by the medical profession. It is also unclear if physicians and the public view patient safety as a major issue. An article published in 2002 showed that neither group ranked medical errors at the top of problems facing medicine. Furthermore, 70% of the public and 25% of physicians thought errors should be reported to a state agency and only 19% of physicians and 46% of the public thought that computerized physician order entry (CPOE) is an effective way to reduce errors. Nevertheless, a survey showed that more patients were concerned about the safety of a prescription than a commercial flight [18]

- **Error reporting.** Reporting continues to be voluntary and inadequate at best. In a recent study of over 90,000 voluntary electronic reports from 26 hospitals, only 2% were reported by physicians, most were from nurses [19]

What is the Federal Government doing about patient safety?

Agency for Healthcare Research and Quality (AHRQ)
The AHRQ is an agency under the Department of Health and Human Services. In 2001 it provided $50 million for research projects to try to reduce medical errors, to include information technology implementation. Importantly, the agency developed the AHRQ Patient Safety Network that includes an e-mail newsletter with medical literature reviews of patient safety articles, found at http://psnet.ahrq.gov . The AHRQ also funds projects to improve healthcare and patient safety through information technology. [20]

Patient Safety and Quality Improvement Act S.544
This patient safety initiative became law July in 2005 and calls for voluntary reporting of medical errors without recrimination. The Department of Health and Human Services will maintain databases to collect data. This act is limited by the fact that there is no definition of "error" and data can only be used in a criminal case and with the physician's permission. Also, it is unclear what the incentive is for the average physician to report errors? [21]

What are the States doing about patient safety?

Florida became the first state to openly report a range of cost and quality measures for both hospitals and outpatient facilities. Two web sites were created. The first is www.FloridaCompareCare.gov that provides broad coverage of data such as infection and mortality rates, in addition to costs for common operations. The second site www.MyFloridaRx.com lists pricing information for the 50 most common drugs prescribed in Florida. [22]

What are private agencies doing about patient safety?

Joint Commission on Accreditation of Healthcare Organizations (JCAHO)
Of the six 2006 National Patient Safety Goals, four have health IT implications
- Improve the accuracy of patient identification: potential for bar coding or RFID
- Improve the effectiveness of communication among caregivers: expedite critical test results to clinicians by using cell phones or voice over Internet protocol (VoiP)

- Improve the safety of using medications: examples could be the use of online or PDA drug programs as well as clinical calculators
- Accurately and completely reconcile medications across the continuum of care: if all pharmacy records were online and linked by a health information exchange it would be easy to reconcile <u>all</u> medications [23]

Institute for Healthcare Improvement (IHI). The IHI instituted a plan in December 2004 to save 100,000 lives from medical errors by getting hospitals to institute at least one of six safety measures. <u>A report on June 14th 2006 estimated that 122,300 deaths have been prevented through the adoption of new safety measures by more than 3,000 participating hospitals over an 18th month period.</u>[24] It should be noted that none of these methods directly involved information technology.

LeapFrog Group
Organization is a consortium of healthcare purchasers that demand better quality. One of the four areas they promote is the adoption of inpatient computerized physician order entry (CPOE). They maintain survey safety data from over 1,000 hospitals as well as a calculator to determine ROI for hospital pay for performance programs. Unfortunately, in 2004 only four percent of hospitals surveyed had fully implemented inpatient CPOE! [25]

HealthGrades
A commercial site that rates different aspects of medical care. Some of the reports are associated with charges ($7.95). Hospital reports compare 28 surgical procedures or diagnoses by state. Physician reports compare disciplinary action, board certification and patient opinions. Medical cost reports compare cost information for 56 common procedures to include doctor, hospital, lab and drug costs and includes out of pocket costs and average health plan payments. [26]

Information Technology related to Patient Safety

Medication error reduction remains the most important patient safety area impacted by healthcare IT. It has been shown that injury from medications (adverse drug events) accounts for up to 41% of hospital admissions and more than $2 billion in inpatient costs.[27] Fortunately, 99% of medication errors do not result in an adverse drug event (ADE). About 30% of ADEs are preventable and of those about 50% are preventable at the ordering stage.[28] In spite of the fact that more drugs are prescribed for outpatients, inpatient drug use is more dangerous. Intravenous (IV) medications are associated with 54% of ADEs and 61% of serious or life threatening errors. [29] Technology has great potential in reducing medication errors, but it is not a panacea. As an example, an article by Oren on technology and medication errors concluded that well controlled studies are lacking, tend to be reported only at a select number of universities and patient outcomes are lacking.[30] Moreover, in a systematic review on the impact of HIT on quality, efficiency and cost reduction the point is made that four institutions are responsible for the majority of what is written on the subject; Brigham and Women's Hospital, Regenstrief Institute, Veterans Administration Hospital system and LDS Hospital/Intermountain Healthcare. [31]

The following technologies have the potential to decrease medication errors:

1. **Computerized Physician Order Entry**
   a. **Inpatient CPOE.** This functionality was recommended by the IOM in 1991. Most studies so far have looked primarily at inpatient CPOE and not ambulatory CPOE. A 1998 study by David Bates in JAMA showed that CPOE can decrease serious inpatient medication errors by 55%.[32] As pointed out in the previous paragraph, many of the studies showing reductions in medication errors by the use of technology were reported out of the same institution. Other hospital systems are unlikely to produce the same optimistic results. With the inception of CPOE we are in fact seeing evidence of new errors that result from technology. A December 2005 article in *Pediatrics* suggested that the mortality rate increased from 2.8% to 6.5% after implementing Cerner's EHR at Children's Hospital of Pittsburgh. They point out however, with the new system they could not use it "until after the patient had physically arrived" and registered in the system. This may have led to delays in diagnosis and treatment. The situation was corrected and we will have to see if mortality rates drop back down to baseline. [33] In another article in the July 2006 journal *Pediatrics* from Children's Hospital and Regional Medical Center in Seattle, they implemented the same EHR and found no increase in mortality. It appears that this was due to better planning and implementation. In this same article Dr Del Beccaro stated that the CPOE system has: eliminated handwriting errors, improved medication turnaround time and helped standardize care. [34] In another article by Nebeker, substantial adverse drug events (ADEs) continued at a VA hospital following the adoption of CPOE that lacked full decision support, such as alerts.[35] There have been other articles in the literature recently that have cast a negative light on EHRs and medication errors. Suffice it to say that clinicians and staff must be properly trained in CPOE; otherwise errors will likely increase, at least in the short term.

   b. **Outpatient CPOE.** Americans made 906.5 million outpatient visits in the year 2000. Although the initial attention was on inpatient CPOE, by shear numbers there is more of a chance for a medication error written for outpatients. According to a report by the Center for Information Technology Leadership, adoption of an ambulatory CPOE system (ACPOE) will likely eliminate about 2.1 million ADEs per year in the USA. This would prevent 1.3 million ADE-related visits, 190,000 hospitalizations and more than 136,000 life threatening ADEs. It is estimated that ambulatory CPOE could save as much as $44 billion/year. [36]

c. **Clinical Decision Support.** Computerized drug alerts have obvious potential in decreasing medication errors but have not been very successful to date. According to a systematic review by Kawamoto et al, successful alerts need to be automatic, integrated with CPOE, require a physician response and make a recommendation. [37] In an interesting study of all alert overrides for 3 months at Brigham and Women's Hospital in Boston they noted that 80% of alerts were over-ridden because: 55% of respondents stated they were "aware"; 33% stated "patient doesn't have this allergy" and 10% stated "patient already taking the medicine". Only six percent of patients experienced ADEs from alert overrides and half were serious. Their conclusion was that alert overrides are common but don't usually result in serious errors. [38] In a newer study at the same institution of outpatient alerts, they found better acceptance when alerts were interruptive only for critical situations. Sixty seven percent of interruptive alerts were accepted which represents an improvement. Many alerts were still incorrect so further improvements are needed. [39] In a third study from Brigham and Women's Hospital, critical lab alerts were automated to call a physician's cell phone. This strategy led to11% quicker treatment and reduced the duration of a dangerous condition by 29%. [40]

d. **Accurate Drug Histories.** The significance of having prior prescribing information available at the time a prescription is written should not be underestimated. Researchers at Henry Ford Health System reported a study where clinicians were given six months of prescription claims data compared to a control group with no such information. Those with the additional information were more likely to change dosages (21% vs 7%); add drugs (42% vs 14%) and discontinue drugs (15% vs 4%). Also, physicians with prior drug histories detected non-compliance in about 1/3 of patients versus none in the control group. [41] Another important issue concerning medication error reduction is the ability to reconcile all outpatient medications when a patient is admitted to a hospital. In many instances the information given by the patient is not correct. Lau reported that 61% of patients had at least one drug missing and 33% had two or more drugs missing on initial admission interview. [42] EHRs, RHIOs and pharmacy claims data all offer the opportunity to provide additional patient drug history.

2. Automated Medication Dispensing Devices
- Kept on nursing units
- Like ATM machines, these devices communicate with pharmacy computers and dispense medications stocked by the pharmacy
- User must have password to access
- Device keeps medication records
- Unfortunately, there is no evidence that these systems reduce errors or affect outcomes, in spite of their high price tag [43]

3. Pharmacy Dispensing Robots
- Studies suggest that robotic systems save space, decrease manpower, increase the speed to fill a prescription and decrease errors
- Very helpful when there is a shortage of pharmacists or staff. Allows pharmacists to have more of a supervisory role
- Ideally, systems would receive electronic prescriptions from outpatient and inpatient areas, then be checked by both the EHR and the pharmacist, then labels are printed and the prescription filled [44]

4. Electronic Medication Administration Record (eMAR)
- Eliminates legibility issues
- No need to rewrite the MAR when medications are changed or discontinued
- No more searching for the patient's chart to see what medications the patient is on
- Can provide allergy and timing alerts
- Available to others, like physicians making rounds
- Can be web based and can be wireless [45]

5. "Smart" Intravenous (IV) Infusion Pumps
- Intravenous sedatives, insulin, anticoagulants and narcotics pose the highest risk of harm from medication errors [46]
- Early pump versions allowed for constant infusion rates without programmable alerts
- Smart pumps can be programmed to deliver the correct amount of IV drugs and are associated with drug libraries and alerts
- Recently Alaris developed a smart pump with built in bar coding [47]

6. Calculators
- Johns Hopkins University created a web based Pediatric total parenteral (IV) nutrition (TPN) calculator and as a result reduced medication errors in half with an annual projected saving of $60-80,000
- The infusion calculator was associated with 83% fewer errors [48]
- Other web based and handheld medical calculators are available but little is known regarding their impact on patient safety

7. Bar Coding

- Can be placed on patient ID bands, medications, vials of blood and transfusion bags. Pictures can be added and in this case a picture of the infant and the mother is present (Fig. 12.2)
- FDA mandated that drug companies apply barcodes on unit dose medications by April 2006. JCAHO is deciding on when to make it mandatory
- Price tag likely $300K- $1 million for hospitals to adopt barcode technology
- There are very few studies looking at patient outcomes with this technology
- One study showed an error rate decrease from 1.0% to .2% and improvement in stock ordering times [49]

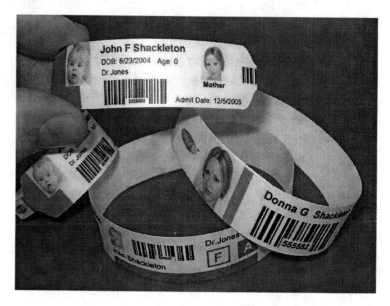

Figure 12.2. Bar coded ID bracelets (courtesy ENDUR ID)

- Veterans Affair hospitals have had bar coding since 1999 in their 161 hospitals. Once scanned the software confirms that the correct medication in the correct dose and frequency has been given to the correct patient. It also updates the electronic medication record. As a result of this technology one VA hospital was able to decrease medication errors by 66% over 5 years [50]
- Only a minority of hospitals use bar coding at this point
- Sutter Health implemented barcode medication administration and found that after one year there were:
    - 28,000 medication mix-ups averted and 9 % could have resulted in moderate to serious harm
    - Cost for bar coding in their 25 hospitals was $25 million [51]

8. Radio Frequency Identification (RFID)

- Unlike bar coding, RFID can be read-only or read-write capable
- Can be read if wet or thru clothing. Better for blood and IV bags
- Drug companies and Walmart want RFID to record and track all inventory. They are also seeking to decrease the counterfeit drug market which may total $2 billion yearly

- Scanner must interface with an established database to identify the object with the RFID tag
- Tags are cheap but transceivers (scanners) are expensive
- Scanners can be part of a PDA or laptop computer. PDA with Socket CF reader-scan card can read RFID and barcodes [52]
- Tags can be active (battery, larger, more memory, longer range and more expensive) or passive (smaller, cheaper, short range and no battery)
- Can be low, medium or high frequency
- Can track patients within a hospital with active tag that works like a transmitter and gives location and time
- RFID tracking will allow for better business and time analysis
- May replace or complement bar coding

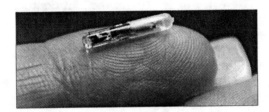

Figure 12.3. Implantable RFID chip (courtesy VeriChip)

Figure 12.3 demonstrates an implantable FDA approved RFID chip has been used in both humans and animals. The chip contains a 16 digit number that is read by a scanner. The number is used to locate the web based medical record. The market may be primarily for infants, the unconscious and the demented. [53]

Conclusion

Better studies are needed before we can expect widespread purchase and implementation of technology to improve patient safety. Until then, we will have to rely on anecdotal and limited studies. Moreover, CEOs and CIOs will be looking for a reasonable return on investment. However, if improved patient safety means a larger market share, fewer law suits or a better hospital ranking by the state or federal government, then adoption will likely occur. According to Healthgrades there is evidence that the highest ranked hospitals for quality have lower mortality rates. [54] Additionally, it appears that the most wired hospitals also have lower mortality rates but it is too early to establish clear cut cause and effect. [55] One could also draw on the experience of the Veterans Affairs hospitals to show how their electronic health record has markedly improved the quality of care and efficiency. [56] Is their dramatic systemic improvement solely due to the universal EHR (Vista CPRS) or is it due to the visionary Dr Kiser who saw the need for modernization and the establishment of a culture of safety?

References

1. Institute of Medicine. Educating Health Professionals to use Informatics http://www.iom.edu/Object.File/Mastehttp://www.iom.edu/Object.File/Master/10/483/Informatics2.pdfr/10/483/Informatics2.pdf (Accessed February 2 2006)
2. Leape LL. Error in medicine. JAMA. 1994 Dec 21;272(23):1851-7.
3. Airline industry since 1970 http://www.airsafe.com/airline.htm (Accessed October 4 2005)
4. Langreth R Fixing Hospitals Forbes June 20 2005
5. World Health Organization http://www.who.int/whr/2004/annex/topic/en/annex_1_en.pdf (Accessed October 4 2005)
6. McGlynn EA et al The Quality of Health Care Delivered to Adults in the United States NEJM 2003:2635-2645
7. To Err is Human: building a safer health system. 2000. IOM http://www.nap.edu/books/0309068371/html/ (Accessed October 10 2005)
8. Modern health care system is the leading cause of death http://www.mercola.com/2004/jul/7/healthcare_death.htm (Accessed October 11 2005)
9. Crossing the Quality Chasm: A New Health System for the 21st Century (2001) Institute of Medicine (IOM) http://lab.nap.edu/books/0309072808/html/3.html (Accessed October 4 2005)
10. Patient Safety: Achieving a new standard of care http://www.nap.edu/catalog/10863.html (Accessed April 19 2006)
11. HealthGrades Quality Study. Second Annual Patient Safety in American Hospital's Report http://www.healthgrades.com/media/DMS/pdf/PatientSafetyInAmericanHospitalsReportFINAL42905Post.pdf (Accessed October 28 2006)
12. Commonwealth Fund Survey. Health Affairs Nov 2005 http://www.cmwf.org/surveys/surveys_show.htm?doc_id=313115 (Accessed September 5 2006)
13. Crossing the Quality Chasm: A New Health System for the 21st Century (2001) Institute of Medicine (IOM) http://lab.nap.edu/books/0309072808/html/3.html (Accessed October 11 2005)
14. http://www.nashp.org/Files/Tues_11_Integrating_Yasnoff.pdf (Accessed November 11 2005)
15. Agency for Healthcare Research and Quality www.ahrq.gov (Accessed October 16 2005)
16. Balas EA, Boren SA., Managing Clinical Knowledge for Health Care Improvement. Yearbook of Medical Informatics 2000.
17. Health Professions Education: A Bridge to Quality (2003) Board on Health Care Services (HCS) Institute of Medicine (IOM) http://darwin.nap.edu/books/0309087236/html/29.html (Accessed February 11 2006)
18. Blendon RJ et al Views of Practicing Physicians and the Public on Medical Errors NEJM 2002;347:1933-40
19. Milch CE et al Voluntary Electronic Reporting of Medical Errors and Adverse Events J of Gen Int Med 2006
20. Agency for Healthcare Research and Quality www.ahrq.gov (Accessed October 16 2005)
21. AHRQ's Patient safety & Quality Healthcare Sept/Oct 2005 56-57
22. Wooten J Gingrich has health care remedies ready. The Atlanta Journal Constitution. September 24 2006 http://ajc.com (Accessed October 1 2006)
23. Joint Commission on Accreditation of Healthcare Organizations www.jointcommission.org (Accessed April 5 2006)
24. Institute for Healthcare Improvement www.ihi.org (Accessed April 5 2006)
25. Leapfrog www.leapfroggroup.org (Accessed April 15 2006)
26. HealthGrades www.healthgrades.com (Accessed December 5 2006)
27. Nebeker J et al High Rates of Adverse Drug Events in a highly computerized hospital Arch Int Med 2005;165:1111-16

28. Bates DW et al Incidence of Adverse Drug Events and Potential Adverse Drug events: Implications for Prevention JAMA 1995;274:29-34
29. Averting Highest Risk Errors Is First Priority. Patient Safety & Quality Healthcare May/June 2005
30. Oren E, Shaffer ER and Guglielmo JB Impact of emerging technologies on medication errors and adverse drug events Am J Health Syst. Pharm 2003;60:1447-1458
31. Chaudhry B et al Systematic Review: Impact of Health Information Technology on Quality, Efficiency and Costs of Medical Care Ann of Int Med 2006;144:E-12-E-22
32. Bates DW et al Effect of computerized physician order entry and a team intervention on prevention of serious medication errors JAMA 1998;280:1311-1316
33. Han YY et al Unexpected increased mortality after implementation of a commercially sold computerized physician order entry system Pediatrics 2005;116:1506-1512
34. Del Beccaro MA et al Computerized Provider Order Entry Implementation: No Association with Increased Mortality Rate in An Intensive Care Unit Pediatrics 2006;118:290-295
35. Nebeker J et al High Rates of Adverse Drug Events in a highly computerized hospital Arch Int Med 2005;165:1111-16
36. Center for Information Technology Leadership. CPOE in Ambulatory Care http://www.citl.org/research/ACPOE.htm (Accessed April 5 2006)
37. Kawamoto K et al Improving clinical practice using clinical decision support systems: a systematic review of trials to identify features critical to success. BMJ 2005 330: 765-772
38. Hsiegh TC et al Characteristics and Consequences of Drug Allergy Alert Overrides in a Computerized Physician Order Entry System JAIMA 2004;11:482-491
39. Shah NR et al Improving Acceptance of Computerized Prescribing Alerts in Ambulatory Care JAMIA 2006;13:5-11
40. Bates DW, Gawande AA Patient Safety: Improving Safety with Information Technology NEJM 2003;348:2526-2534
41. Bieszk N et al Detection of medication non-adherence through review of pharmacy claims data Am J Health Syst Pharm 2003; 60:360-366
42. Lau HS et al The completeness of medication histories in hospital medical records of patients admitted to general internal medicine wards Br J Clin Pharm 2000;49:597-603
43. Murray M Automated Medication Dispensing Devices http://www.ahrq.gov/clinic/ptsafety/chap11.htm (Accessed April 23 2006)
44. Hospital Pharmacist http://www.pjonline.com/pdf/papers/pj_20050618_automateddispensing.pdf (Accessed April 10 2006)
45. Ascend eMAR www.hosinc.com (Accessed April 23 2006)
46. Winterstein AG, Hatton RC, Gonzalez-Rothi R Identifying clinically significant preventable adverse drug events through a hospital's database of adverse drug reaction reports. Am J of Health Sys Pharm 2002;59:1742-1749
47. Vanderveen T IVs First, a New Barcode Implementation Strategy Patient Safety & Quality Healthcare May/June 2006
48. Ball MJ, Merryman T, Lehmann CU. Patient Safety: A tale of two institutions. J of Health Info Man 2006;20:26-34
49. Chester M, Zilz D Effects of bar coding on a pharmacy stock replenishment system. Am J Hosp Pharm 1989;46:1380-5
50. Coyle GA, Heinen M Evolution of BCMA within the Department of Veterans Affairs. Nurs Admin Q 2005;29:32-38
51. Bar codes help Sutter avoid medication errors November 24 2004 www.ihealthbeat.org (Accessed October 7 2005)
52. Iglesby T Ready for Prime Time? Patient Safety & Quality Healthcare May/June 2006 p.50-5
53. Verichip http://www.4verichip.com/nws_10132004FDA.htm (Accessed April 5 2006)
54. Study: Hospitals rated in top 5% have mortality rates 27% lower. Patient Safety & Quality Healthcare March/April 2006: 57
55. Annual list of most-wired hospitals released. 7/12/2005. www.ihealthbeat.org (Accessed April 5 2006)

56. Stires, D Technology has transformed the VA CNN Money http://money.cnn.com 5/25/2006 (Accessed May 5 2006)

Chapter 13: Electronic Prescribing

Learning Objectives

After reading this chapter the reader should be able to:

- List the potential benefits of electronic prescribing
- Identify the problems and limitations of handwritten prescriptions
- Describe the SureScripts and RxHub networks
- List the obstacles to widespread e-prescribing

Introduction

The prescription shown in figure 13.1 resulted in a patient death as the pharmacist interpreted the drug prescribed as *Plendil* and not *Isordil*. This was the first medical malpractice case successfully prosecuted due to illegible handwriting.[1]

Cases such as this led the Institute for Safe Medication Practices (ISMP) to push to "eliminate handwritten prescriptions within three years". The ISMP points out that up to 7,000 Americans die each year due to medication errors resulting in about $77 billion annually.[2] In 2006 the Institute of Medicine stated that all prescriptions should be electronic by the year 2010.[3] Since 2000 Delaware, Florida, Idaho, Washington, Montana, Tennessee, and Maryland have enacted laws requiring legible prescriptions. Montana fines up to $500 for each illegible script. Washington State took a further step by outlawing prescriptions written in cursive.[4]

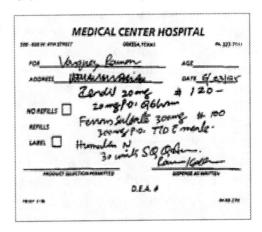

Figure 13.1. Illegible prescription

Approximately 3 billion prescriptions are written yearly in the United States and of those only about 2-3% are electronic. On the surface this seems difficult to understand given the multiple advantages of e-prescribing:

- Legible, complete prescriptions eliminating handwriting errors and decreasing pharmacy "callbacks" (150 million yearly) and rejected scripts (30%) [5]
- Not only is prescription legible, abbreviations and unclear decimal points are avoided
- Time to pick up prescription would be reduced due to electronic transmission

- Fewer duplicated prescriptions
- Timely notification of drug alerts and updates
- Better use of generic or preferred drugs
- Interfaces with practice and drug management software
- Secure and HIPAA compliant process
- Ability to look up drug history, drug-drug interactions and compliance
- Ability to check eligibility, co-pays and file drug insurance claim

The Center for Information Technology Leadership's 2003 Report on the Value of Computerized Physician Order Entry in Ambulatory Settings estimated that e-prescribing would save $29 billion annually from fewer medication errors; reduced overuse, misuse and adverse drug event related hospitalizations and more cost effective selection of generic or less expensive medications. [6]

The concept of e-prescribing is gathering momentum in the United States. Alaska and West Virginia are the last states two states to approve eRx and they are in the process of approval. [7] Perhaps one of the strongest driving forces behind e-prescribing is the Medicare Prescription Drug Improvement and Modernization Act (MMA) of 2003 that allocates about $50 million in 2007 to support e-prescribing systems and allows health plans to offer incentives to adopt information technology. Congress also approved the exceptions that allow for donations of e-prescribing technologies to physicians from hospitals and other entities. Initial standards for e-prescribing were in place by January 2006 when the Medicare prescription benefit began. Final standards should be finalized by 2008 or 2009.[8] Under this new Act, prescription drug plans will be required to offer e-prescribing, although the option is voluntary for physicians. This program has tremendous clout as 40% of all scripts written are covered by Medicare. The database created for this program would be the largest ever related to prescribing. The Act will also offer drug plans higher reimbursement for physicians who e-prescribe. [9]

New electronic standards for e-prescribing were released just prior to the January 2006 deadline:
- National Council for Prescription Drug Programs (NCPDP) Script version 5 will be required for new scripts, refills and changes. This standard is administrative in nature
- Accredited Standards Committee (ASC) X12N 270/271 version 4010 will be required for benefit questions between prescribers and Medicare part D sponsors
- NCPDP Telecommunication Standard Version 5.1 required for eligibility questions between dispensers and Medicare sponsors [10]

As is the case with most emerging technologies, some companies have folded and others have consolidated. Table 13.1 includes most of the significant current software vendors of e-prescribing.

Vendor	Web Address
Allscripts eRx	http://www.allscripts.com/products/physicians-practice/eprescribing-med-services/erxnow/default.asp
DrFirst	http://www.drfirst.com/index.jsp
iScribe	http://www.iscribe.com
SynaMed	http://www.synamed.com
PocketScript	http://www.zixcorp.com/
Purkinje	http://www.purkinje.com/en/solutions17.cfm
RxNT	http://www.rxnt.com/

Table 13.1. E-prescribing vendors

E-prescribing can be a standalone software program in a PDA, PC, tablet PC or be web based. In addition, it can also be an integrated into an electronic health record. To date, not all EHR vendors offer this functionality. In order for e-prescribing to function well, patient lists need to be uploaded into the system so the clinician doesn't have to waste time typing patient information each time a script is submitted. Importantly, e-prescribing networks have been created that would allow for new prescriptions and renewals to be sent electronically to the patient's pharmacy of choice.

SureScripts was founded in 2001 by the National Association of Chain Drug Stores (NACDS) and the National Community Pharmacists Association (NCPA) to improve the quality, safety and efficiency of the overall prescribing process. It seems likely that one of the strongest motivating forces behind this collaboration was the need to reduce the number of call backs by pharmacists, as they reduce the pharmacist's bottom line. The SureScripts Electronic Prescribing Network is the largest network to link electronic communications between pharmacies and physicians, allowing the electronic exchange of prescription information. By 2006 more than 90% of the 55,000 retail pharmacies in the U.S. signed letters to become members of SureScripts but only 50% were actually connected at that point. It is not known how many physicians have connectivity. SureScripts works with software companies that supply electronic health records (EHRs) and electronic prescribing applications to physician practices and pharmacy technology vendors to connect their solutions to the SureScripts Electronic Prescribing Network. The network is free to physicians but pharmacies pay a small amount to the software vendor like DrFirst and they in turn pay SureScripts. Vendors cannot connect to the SureScripts Electronic Prescribing Network until they complete a certification process that establishes rules that safeguard the prescribing process, to include patient choice of pharmacy and physician choice of therapy. SureScripts does not sell, develop, or endorse software.[11] Vendors are certified based on several services: e-prescribing, e-refills, Rx history, eligibility and formulary. Many vendors are certified only for e-prescribing and e-refills at this point. Drug history data comes from prior prescriptions and not insurance claims information. A list of the certified software companies and EHR vendors that are connection capable is available.[12]

RxHub was created in 2001 by three of the leading pharmacy benefit manager (PBM) organizations: AdvancePCS (acquired by Caremark), Express Scripts, and Medco Health Solutions. These organizations are responsible for funding and administering drugs on behalf of insurance companies and employers in order to control costs. RxHub creates another network between physicians and pharmacies to route patient medication histories (based on claims data) and pharmacy benefit information to physicians. This helps determine if the patient is eligible to receive certain drugs based on the insurance plan. Figure 13.2 shows the two different networks. This network also provides the option of forwarding a prescription to one of the mail order pharmacies. RxHub MEDS is a separate program that makes outpatient prescription histories available for inpatient care.[13]

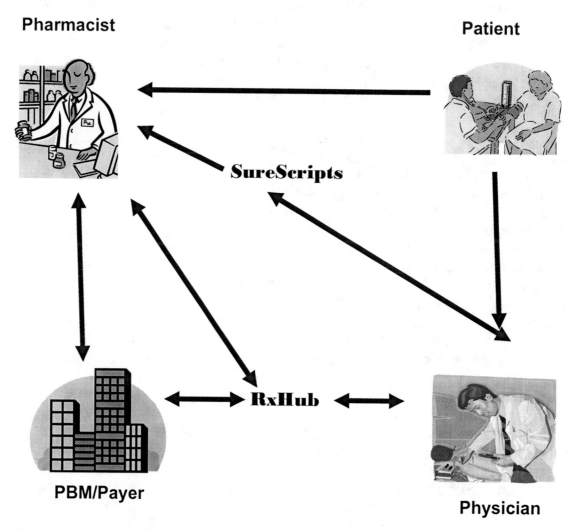

Figure 13.2. SureScripts and RxHub networks

E-Prescribing Studies and Projects

Annals of Family Medicine looked at prescribing habits of 19 physicians using e-prescribing compared to a control group. Physicians first selected the likely diagnosis and they were then provided a list of recommended choices (clinical decision support) as well as links to evidence based resources. Drug costs were reduced by 11% compared to controls; but note that several of the authors were affiliated with one of the vendors.[14]

Oregon Health and Science University compared handwritten versus computerized prescriptions in an emergency room setting. Computer written scripts were three times less likely to include errors and five times less likely to require a pharmacist's clarification. This resulted in a decrease in wait time for patients and call back time by pharmacists to physicians.[15]

E-Script Pilot Study studied 100 physicians in the Washington, DC area using DrFirst software in 2005. 0.3% scripts generated serious drug interactions or allergy alerts. The study estimated an annual avoidance cost of $100,000 and an average savings to health plans of $29 per prescription filled due to improved use of generics and drugs of choice.[16]

Maryland Safety through Electronic Prescribing Initiative created a consortium of 27 health organizations. The goal was to expedite the adoption of electronic prescribing in order to reduce medication errors. They will use multiple training modalities such as conferences, workshops, etc.[17]

Wellpoint is the leading health plan in the United States covering 34 million members. They offered physicians either a free office automation package or a free e-prescribing system using Dell Axium PDA's. Only 2,700 physicians out of the 25,000 contacted signed up for the e-prescribing alternative. This outcome was a reminder that it is not enough to make a service free. You must provide training and physicians have to believe that in the long run it will save time or money for them or their patients.[18]

South East Michigan E-Prescribing Initiative. Three major auto makers have partnered with the three largest health insurers to promote e-prescribing. The project will use e-prescribing software including RxHub, hosted on PCs or PDAs. The goal is to sign on 17,000 physicians eventually. Hardware was not provided and it is unclear who will pay for the software. Results thus far have shown increased use of generic drugs resulting in a $3.1 million savings and 80,000 prescriptions changed or cancelled due to alerts.[19]

Nevada Project. Sierra Health Services and Clark County Medical Society collaborated to provide Allscripts e-prescribing software for its 5,000 physicians, for free, for 10 years. Maintenance fees will be waived for two years. According to one report, utilization of generic drugs has increased from 53% to 64% (estimated cost saving of $5 million). Although they state that medical errors are down, specifics were not provided.[20]

Tuft's Health Plan Experience 2001-2002. The study involved 226 network physicians and allied health providers. As a result of the initiative, 8.9 fewer safety errors per physician per year were reported. Cost saving of $.30-.40 PMPM (per member per month) due to use of generics and preferred drugs was also noted. Clinicians and office staff reported that e-prescribing saved as much as 2 hours per day for the doctor's office and up to one hour per day for the pharmacist. [22]

Massachusett's e-Rx Collaborative. Program was formed by Blue Cross Blue Shield, Tufts Health Plan and the Neighborhood Health Plan. Two e-prescribing vendors (Zix and DrFirst) provided a PDA or PC based program to about 3000 physicians, paid for by the Collaborative. After the trial runs out in 2006 it will cost about $600/year/clinician. Software will work with a Pocket PC PDA using PocketScript or with a web based program on the desktop. The system can operate wired or wirelessly and work with a PDA or a BlackBerry. [22]

Florida Medicaid has awarded a contract to drug reference company Gold Standard to provide PDAs with drug information. The program began in 2001 with 500 users with the goal of expansion to 3000 physicians. The program and PDAs are free to clinicians and features wireless technology. A major goal is to promote use of more cost effective drugs using a preferred formulary. It is important to note that the average cost for a one month supply of a generic drug is $22.79, whereas the average for a brand name drug is $77.29.[23] As a result, 39 states require a generic drug for Medicaid patients when available. Florida is the first state to provide PDAs at no cost to try to reduce Medicaid drug costs. It is anticipated that the program will pay for itself by approximately 2006.[24-25]

Horizon Blue Cross/Blue Shield of New Jersey has partnered with three e-prescribing vendors to deliver eRx software to 665 physicians. To date, they have written over 400,000 prescriptions at a cost of over $5 million dollars to Horizon Blue Cross Blue Shield. Results so far show a savings of 30-60 minutes daily. [26]

National E-prescribing Patient Safety Initiative (NEPSI) was announced on January 15th 2007 and represents the most significant e-prescribing initiative thus far. The goal is to supply free web based e-prescribing software (eRx Now™) to every physician in the United States. The two main sponsors are Allscripts and Dell Computers. Other coalition partners include Google, Aetna, Cisco, Fujitsu, Intel, Microsoft, Sprint, WellPoint, Wolters Kluwer Health and others. It is stated that the program will include important features like drug-drug information, benefit and formulary status information and only require 15 minutes of training. The system will use SureScripts to transmit information from physicians to pharmacies. National deployment will begin in incremental steps in March 2007. Preliminary information is available at these sites. [27-28] Questions still remain in terms of the degree of commercial influence, whether data mining will occur, why RxHub in not a partner and how the software program will auto-populate patient demographics so a physician doesn't have to type in rows of patient data every time he/she wants to e-prescribe. This project will test how ready SureScripts really is as well as clinicians who have been sitting on the sidelines.

E-Prescribing Obstacles and Issues

- Adoption rate is low. According to a 2004 study, only 14% of physicians use eRx and 62% are in group practice [29]
- Who will pick up the tab? Payers? Physicians? Consumers? Government? At this time physicians are reluctant to pay because they believe others such as pharmacists and PBMs benefit more from eRx. Implementation will also require training and upgrades of office technology to accommodate wireless transmission and other issues such as uploading patient demographics. Physicians don't want to lose time learning to use a new system if there is no reimbursement or waste time because eRx doesn't integrate with other office IT systems. With the new NEPSI web based initiative, cost is less of a factor, so time will tell if this will launch e-prescribing more rapidly
- The 2003 Medicare bill and the new JCAHO safety goal of reconciling medications will promulgate the use of eRx
- Federal grants will be necessary to test electronic prescribing standards prior to final adoption [30]
- E-prescribing or eRx will eventually be part of all EHRs. In 2007 the Certification Commission for Healthcare Information Technology will only require that a prescription can be sent to a pharmacy and that a response back is received [31]
- Some pay for performance programs reimburse for eRx adoption [32]
- E-prescribing is slower than paper scripts, but not when you factor in time spent calling back pharmacists or playing "phone tag"
- Most authorities at this time have an attitude of cautious optimism [33]

Conclusion

E-prescribing as well as electronic networks between pharmacies and clinician's offices are now a reality. Evidence so far indicates that e-prescribing should save time and money and hopefully result in improved patient safety. We will watch the new NEPSI initiative with great anticipation to see if this provides the necessary impetus to make e-prescribing the standard of care.

References

1. Poison in Prescription. Allscripts 3/19/2001 http://www.allscripts.com/ahsArticle.aspx?id=297&type=News%20Article (Accessed November 2 2005)
2. A call to Action: Eliminate Handwritten Prescriptions Within 3 years! White Paper. Institute for Safe Medication Practices www.ismp.org (Accessed November 3 2005)
3. IOM Report Calls for E-prescribing www.ihealthbeat.org July 21 2006 (Accessed July 24 2006)
4. State Initiatives to Avoid Prescription Drug Errors December 2006 www.ncsl.org/programs/health/rxerrors.htm (Accessed February 22 2007)
5. RxHub www.rxhub.net (Accessed February 22 2007)
6. The Value of Computerized Provider Order Entry in Ambulatory Settings 2003 www.citl.org/research/ACPOE.htm (Accessed February 22 2007)
7. SureScipts www.surescripts.com (Accessed February 22 2007)
8. Medicare Modernization Act www.cms.hhs.gov/medicarereform/ (Accessed November 8 2005)
9. Sarasohn-Kahn. Medicare's broadening shoulders www.ihealthbeat.org July 29 2005 (Accessed July 30 2005)
10. www.healthcareitnews.com Nov 8 2005 (Accessed November 8 2005)
11. SureScripts www.surescripts.com (Accessed December 18 2006)
12. Surescripts http://www.surescripts.com/get-practice-connected.htm (Accessed November 8 2005)
13. RxHub www.rxhub.net (Accessed February 8 2007)
14. McMullin ST Impact of an Evidence-Based Computerized Decision Support System on Primary Care Prescription Costs Ann of Fam Med 2004;2:494-498
15. Bizovi KE et al The Effect of Computer-Assisted Prescription Writing on Emergency Department Prescription Errors Acad Emerg Med 2002;9:1168-1175
16. E-Script Pilot Study www.healthdatamanagement Feb 7 2005 (Accessed February 10 2005)
17. Maryland Consortium to advance e-prescribing effort www.ihealthbeat.org 2/09/2006 (Accessed February 10 2006)
18. www.healthcareitnews.com April 25 2005 (Accessed April 28 2005)
19. Automakers' E-prescribing program reduces errors, costs www.ihealthbeat.org 2/23/2006 (Accessed February 25 2006)
20. Ackerman K Nevada Physicians Use E-Prescribing to Reduce Medial Errors www.ihealthbeat.org 10/14/05 (November 20 2005)
21. Getting Physicians Connected http://www.tuftshealthplan.com/pdf/epresribing_results_summary.pdf (Accessed July 10 2006)
22. Kowalczyk L Take one and stop scribbling on pads of paper www.Boston.com Jan 10 2005 (Accessed January 12 2005)
23. Generic Pharmaceutical Association www.gphaonline.com (Accessed April 3 2006)
24. Florida Medicaid Expands PDA, E-prescribing Software Program www.ihealthbeat.org 8/12/04. (Accessed August 13 2004)
25. More Florida Docs Could Get Handheld Prescribing Devices www.ihealthbeat.org 5/06/05 (Accessed May 7 2005)
26. http://www.horizon-bcbsnj.com/newsroom/news_releases.asp?article_id=614&urlsection (Accessed November 8 2005)
27. NEPSI www.nationalerx.com (Accessed January 18 2007)
28. eRxNow. Allscripts. www.allscripts.com/products/physicians-practice/eprescribing-med-servicesk/erxnow/default.asp (Accessed January 18 2007)

29. Manhattan Research. Taking the Pulse v. 5.0: Physicians and Emerging Information Technologies Dec 7 2004 www.manhattanresearch.com (Accessed December 10 2004)
30. Pilot Project Launched to Expand Electronic Prescribing. Patient Safety and Quality Healthcare. www.psqh.com 3/01/06
31. CCHIT www.cchit.org (Accessed March 3 2007)
32. Sarasohn-Kahn J, Holt M The Prescription Infrastructure: Are we ready for ePrescribing? California Healthcare Foundation Jan 2006 www.Chcf.org (Accessed January 25 2006)
33. Miller R et al Clinical Decision support and electronic prescribing systems: at time for responsible thought and action JAMIA 2005;12:403-9

Chapter 14: Telemedicine

Learning Objectives

After reading this chapter the reader should be able to:
- State the difference between telehealth and telemedicine
- List the various types of telemedicine such as teleradiology and teleneurology
- List the potential benefits of telemedicine to patients and clinicians
- Identify the different means of transferring information with telemedicine such as store and forward
- Describe the concepts of home and hospital telemonitoring
- Enumerate the most significant ongoing telemedicine projects

Introduction

According to the Office for the Advancement of Telehealth (OAT), Telehealth is defined as:

> "the use of electronic information and telecommunications technologies to support long-distance clinical health care, patient and professional health-related education, public health and health administration" [1]

Similar to the term e-health, telehealth is an extremely broad term. One could argue that Regional Healthcare Information Systems (RHIOs), Picture Archiving and Communication Systems (PACS) and e-prescribing are also examples of telehealth if they exchange healthcare information between distant sites. Some authorities use telehealth as the broader term that incorporates clinical and administrative transfer of information, whereas telemedicine relates to remote transfer of only clinical information. In this chapter we will use the term telemedicine instead of telehealth. Telemedicine can be defined as:

> "the use of medical information exchanged from one site to another via electronic communications to improve patients' health status". [2]

There are three major types of telemedicine in terms of how the information is transferred:
- Store-and-forward. As an example, a primary care physician takes a picture of a rash with a digital camera and forwards it to a specialist

to view when time permits. Commonly used for specialties like Dermatology and Radiology

- <u>Real time</u>. A specialist at a medical center views video images transmitted from a remote site and discusses the case with a physician. This requires more sophisticated equipment to send images real time and often involves two way interactive televisions. Telemedicine also enables the sharing of images from peripheral devices such as stethoscopes, otoscopes, etc

- <u>Remote monitoring</u>. A technique to monitor patients at home, in a nursing home or in a hospital for personal health information or disease management

Telemedicine can also be subdivided into the following categories:
- Traditional Telemedicine: Teleradiology, Teledermatology, Telecardiology, Telepathology, Telesurgery, Telepsychiatry Teleneurology, Teleophthalmology, Telepharmacy, etc
- Telerounding of inpatients
- Telemonitoring and televisits of patients at home
- Telemanagement of patients

Traditional Telemedicine

Currently there are over 130 telemedicine programs that are operational in 48 states.[3] Most programs consist of a central medical hub and several rural spokes. Programs attempt to improve access to services in rural and underserved areas, to include prisons. This reduces travel time and lowers cost for specialists and patients alike. Programs have the potential to raise the quality of care delivered and help educate remote rural physicians. The most commonly delivered services are mental health, dermatology, cardiology and orthopedics. Telemedicine can also be found in the international boating world where sailors can access a medical resource site. After registration they can call, fax or e-mail the site for advice on medical treatment while at sea.[4] Similarly, Virgin Atlantic airlines will equip all of its aircraft with telemedicine devices for emergencies by the year 2009. Satellite technology will transmit the patient's vital signs to MedAire Centre in Arizona for interpretation by medical experts.[5]

- *Teleradiology*. The military has taken the lead in this area partly due to the high attrition rate of Radiologists. By the year 2007 most Army x-rays will be digital, which will help the storage, transmission and interpretation of images. A computerized tomography (CT) scan performed in Afghanistan can be read at the Army medical center in Landstuhl, Germany. The US Army uses a small single image server for picture archiving and communication by the vendor MedWeb.[6] Another example of military teleradiology can be found on the Navy hospital ships Mercy and Comfort where digital images can be transmitted to shore based medical centers. In the civilian sector, NightHawk Radiology Services help smaller hospitals by supplying "offshore" Radiology services stationed in Switzerland and Australia. All are board certified; most trained in the United States and carry multiple state licenses. The turn around time for an image to be read is less than 30 minutes with an average cost of $55.[7] Another more routine but important example of teleradiology is the practice of radiologists reading films from home. They must have high resolution monitors and high speed connections to the Internet but with this set up and voice recognition software, they can be highly productive at home. This is becoming the standard practice for radiologists, instead of driving in or staying at the hospital at night to interpret images.

- **Telesurgery.** The initial approach was to "telementor" surgeons performing operations in remote sites. In 2001 surgeons in New York were able to successfully perform laparoscopic cholecystectomies (gallbladder removal) on six pigs located in Strasbourg, France. This was followed by the uneventful remote removal of the gallbladder in a 68 year old woman; the first case of telesurgery in a human [8]

- **Teleneurology.** Many regions lack neurologists to see patients with stroke like symptoms to determine if they need clot busting drugs (thrombolytics) or need to be transferred to a higher level center. With the advent of telemedicine, the case can be discussed real time and the patient and their x-rays can be viewed remotely by a stroke specialist. One company REACHMD Consult has developed a web based solution that includes a complete audio-visual package so Neurologists can view the patient and their head CT (CAT scan). REACHMD Consult was developed by Neurologists at the Medical College of Georgia. Because the program is web based, the physician can access the images from home or from the office. Likewise, the referring hospital only has to have an off-the-shelf web camera, a computer and a broadband Internet connection [9]

- **Telepharmacy.** Like teleradiology, this field arose because of the shortage of pharmacists to review prescriptions. Vendors now sell systems with video cameras to allow pharmacists to approve prescriptions from a remote location. This is very important at small medical facilities or after-hours or on weekends when there is not a pharmacist on location [10]

Telerounding

This is a new concept developed to help solve the shortage of physicians and nurses. Telerounding is being rolled out in facilities with reasonably good reviews in spite of obvious criticisms that it further compromises the already strained doctor-patient relationship.

- **Robot Rounds.** A study in 2005 in the Journal of the American Medical Association showed that surgeons could make a second set of rounds using a video camera at the patient's bedside (InTouch Robots). A physician assistant makes the actual rounds, backed up by the attending physician remotely via the robot. Robot units are 5 ½ feet tall, weigh 220 lbs and have a computer monitor as a head. The cost is more than $100,000 each or they can be leased for $5000 monthly plus $5000 per viewing station. At this time they are being used in 20 plus hospital systems in the United States. They can move around and can project x-ray results to the patient. Physician satisfaction has been high but a study at Johns Hopkins showed that only 57% of patients were comfortable with continued robotic care [11-13]

- *E-ICU Rounding.* In the United States it is predicted that we need approximately 35,000 Intensivists (physicians who specialize in ICU care), but we have only 6,000. Therefore, remote monitoring makes sense particularly during nighttime hours when physicians might not be present. The Leapfrog Group has advocated care delivered by Intensivists for all ICUs as one of its four patient safety recommendations.[14] Hospitals that use e-ICUs tout the patient safety aspect but the financial savings may be just as significant. An e-ICU service will be less expensive than recruiting full time Intensivists. Avoiding law suits in the ICU also means a cost saving. In a study in 2004, monitors were placed in two large ICUs serving 2000+ patients over 2 years. ICU and hospital mortality and length of stay were compared before and after intervention. Patients were monitored remotely from 12 pm-7 am. The results show that mortality was 9.4% compared to 12.9% for conventional care. Length of stay was 3.63 days compared to 4.35 days for conventional care. It is estimated that over 100 hospitals now have e-ICU programs, even though there is not specific reimbursement.[15] The leading vendor in this area is VISICU with about 150 customers. Sentara Healthcare System in Norfolk, Virginia believes they saved $2,150 per patient using e-ICUs. [16] VISICU extended support of care outside the ICU in December 2006. Their plan now is to use the eCareMobileTM unit to monitor sick patients on medical surgical floors, emergency departments, step down units and post anesthesia units. At Parkview Health in Fort Wayne Indiana, e-ICU physicians will assist rapid response teams that attend to deteriorating patients. Thus far, there has been a 32% reduction in floor based cardiac arrests and improved nursing satisfaction [17]

Telemonitoring

The first home monitoring unit was created just 11 years ago but the concept goes back to circa 1924.[18] At least 55 companies offer technology to monitor patients at home. Vital signs, weights, blood sugars, etc can be sent wirelessly from homes to physician's offices and databases. There are multiple reasons telemonitoring may catch on:

- Medicare changed reimbursement to home health agencies from the number of visits to a diagnoses based system, leading to decreased reimbursement for visiting nurses

- Telemonitoring programs allow for audio and visual contact with patients at home and therefore they can save a visit by a nurse or physician. Instead of nurses making routine home visits, then can make visits only if there is a problem such as a change in symptoms or vital signs

- One consulting organization predicts a nursing shortage of 800,000 and a physician shortage of 85,000 to 200,000 by the year 2020.[19]

- When "baby boomers" turn 65 they will be tech savvy and more likely to demand services like telemonitoring

- Monitoring may be possible also using the ubiquitous cell phone

- Miniaturized biosensors may change the way we view telemonitoring

- Chronic illnesses are on the rise and will likely increase hospitalizations and readmissions so any measure like home monitoring might decrease admissions. The goal is to intervene immediately rather than wait till the next appointment

- Linking home monitoring devices to EHRs and decision support will increase the functionality

- The potential to save costs is attractive but will require high quality confirmatory studies

Health Buddy is an example of a popular home monitoring system with the following features:

- A FDA approved device that is certified by the National Committee for Quality Assurance

- Data is sent via phone lines
- Device comes with desktop decision support software
- Could be part of a disease management program as it covers 45 disease protocols
- Connects to a glucometer, BP machine, weight scales and peak flow meter for asthmatics
- Interactive with patients
- Currently used by over 12,000 patients and has been shown to improve clinical outcomes including reduced hospital admissions, reduced inpatient bed days, increased medication compliance and reduced costs [20]
- Centers for Medicare and Medicaid Services will test the system:
  - CMS will provide the system for 2,000 patients with chronic diseases in Oregon and Washington State
  - Goal is to reduce medical costs by 5% [21]

HoneyWell HomMed
- Provides voice messages to patients in multiple languages
- Standard and optional features: digital weight scale, blood pressure, oximetry, glucometer, peak flow meter, blood tests (PT/INR), temperature and EKG
- Data transmitted via phone lines
- They have over 15,000 monitors currently in use and more than 300,000 patients have been monitored
- Monitor weighs < 3 lbs
- Cost is $3500 for monitoring system [22]

Telemanagement

More and more companies are developing home monitors and sensors that will transmit more information to a physician's office or other healthcare organization. They perceive more of a need based on our graying population, more chronic disease and expensive home care. Programs will be interactive and include patient education for subjects such as drug compliance. This data may interface with an electronic health record or information system or a web site for others to evaluate. Some predict that houses will be wired with multiple small sensors known as "motes" that can monitor daily activities such as taking medications and leaving the house. The information would be transmitted to a central organization that would notify the patient and/or family if there was non-compliance or a worrisome trend. At this point the patient or families will pay for the systems. Some have already complained about the perceived lack of privacy and the potential for too many alerts or alarms [23-25]

Telemedicine Projects

- Informatics for Diabetes Education and Telemedicine (IDEATel) is the largest government sponsored telemedicine program in the US. The project will compare approximately 2000 telemedicine patients with those receiving only normal care in urban and upstate New York State. Patients can maintain an active log of their blood sugars and blood pressure. It may make patients devote more time each day taking care of one of their chronic diseases, particularly if a nurse frequently inquires about their status and lab results. Project will be extended to 2007 [26]

- Indian Telemedicine Project: India plans to launch a satellite to be used exclusively for telemedicine. This is because 75% of Indians live in rural communities, whereas 75% of physicians practice in urban areas. Primary purpose is better healthcare for rural communities [27]

- Telemedicine for Pakistan Earthquake: Virginia Commonwealth University will establish 13 telemedicine stations to aid earthquake relief efforts in Pakistan. Remote areas will be linked to Pakistani hospitals and equipment will be paid for by grants [28]

- Telemonitoring using cell phones: Nurses took pictures of leg ulcers and wounds and then e-mailed the pictures to physicians at the University Hospital of Geneva, Switzerland. Quality of pictures was adequate in the majority of cases. Cell phones proved to be more cost effective than more elaborate telemedicine systems [29]

- Georgia Telemedicine Network: First state-wide effort to link 36 rural hospitals and clinics with specialists at large urban hospitals. Partnership made between Blue Cross/Shield and state government. Telemedicine consults were reimbursed as office visits and 20 specialties were felt to be appropriate for telemedicine [30]

- University of Texas Medical Branch at Galveston: Program is the largest telemedicine system in the world with 300 locations and 60,000 annual telemedicine sessions. Sixty per cent of visits deal with a prison population. The also offer specialty service in neurology, addiction medicine and psychiatry [31]

- VA Rocky Mountain Healthcare Network: In Colorado veterans with heart failure, diabetes and emphysema were enrolled in a telemedicine program. So far, the VA reports there is a 53% reduction in hospital stays resulting in a $508,000 savings overall. Outpatient visits have dropped 52% and overall estimated savings of the program was $1.2 million dollars [32]

- Intel Telemedicine: Project will provide the real time video technology to service 105 rural clinics and hospitals for the municipal government of Zhanjiang, China [33]

- Teleburn Project: University of Utah Burn Center used telemedicine to treat burn patients in three states. Specialists can view videos or digital photos of burn patients for initial determination or follow up. Demonstration project was funded by the Department of Commerce [34]

- Telepsychiatry: Eighteen states cover telemedicine under their Medicaid programs and eight specifically cover telepsychiatry. It is unknown how successful this approach has been however [35]

- Veterans Teleretinal Program: Since 2000 the Veterans Health Administration (VHA) has run a Teleretinal Imaging project at 104 sites to conduct retinal screening for diabetic damage in the 20% of veterans with diabetes. Since all VA clinics do not have retinal experts, the high resolution retinal images are stored and forwarded to ophthalmologists for interpretation [36]

- Federal Communications Commission (FCC): In 2006 they initiated a $400 million yearly budget for pilot projects to promote broadband networks in rural areas. The goal is to create networks for public healthcare organizations and non-profit clinicians that will eventually connect to a national backbone (Internet 2). The network could be used for telemedicine or other medical functions in rural areas [37]

- USDA Rural Development Telecommunications Program. The USDA has a program to finance the rural telecommunications infrastructure. In 2007 there will be grants and loans totaling $128 million to achieve the goals of broadband access for distant learning and remote medical care. [38]

Telehealth Organizations
- Office for the Advancement of Telehealth: falls under Health Resources and Services Administration (HRSA) which is an agency of the Department of Health and Human Services. Its goal is to promote telemedicine in rural/underserved populations, provide grants, technical assistance and "best practices" [1]

- American Telemedicine Association: a non-profit international organization with paid membership that began in 1993. Goals of the ATA are as follows:
    - "Educating government about telemedicine as an essential component in the delivery of modern medical care
    - Serving as a clearinghouse for telemedical information and services
    - Fostering networking and collaboration among interests in medicine and technology
    - Promoting research and education including the sponsorship of scientific educational meetings and the *Telemedicine and e-Health Journal*
    - Spearheading the development of appropriate clinical and industry policies and standards" [2]

Barriers to Telemedicine

- Limited reimbursement. Many private insurers don't cover telemedicine, but a few mandate the same coverage as a face to face visit
- Slow clinical acceptance
- High cost or limited availability of high speed telecommunications
- Bandwidth issues, particularly in rural areas where telemedicine is most needed
- State licensure laws when telemedicine crosses state borders. Some states require participating physicians to have the same state license
- Lack of standards
- Lack of evaluation by a certifying organization
- Fear of malpractice
- Sustainability due to inadequate business plan

Conclusion

Telemedicine is still in its infancy in most areas of the country. The barriers are largely financial due to the cost to set up the system and the lack of reimbursement in most cases. Governmental support is important but Medicare and Medicaid only covers some aspects of telemedicine.[39] Fortunately, the price of Telemedicine systems is dropping so it may become cheaper to have Telemedicine in rural areas than to refer patients to distant specialists. Lastly, if the FCC initiative is successful or RHIOs flourish, then telemedicine may become routine throughout the United States.

References

1. Office for the Advancement of Telelhealth http://www.hrsa.gov/telehealth/ (Accessed December 1 2006)

2. American Telemedicine Association
    http://www.atmeda.org/about/aboutata.htm (Accessed September 10 2006)
3. Puskin DS HHS Perspective on US Telehealth
    www.ieeeusa.org/volunteers/committees/mtpc/Saint2001puskin.ppt
4. Jacobs M. Telemedicine. Sail Magazine June 2006. pp. 60-61
5. Virgin Atlantic to Equip Airplanes with Telemedicine devices.
    www.ihealthbeat.org May 24 2006 (Accessed May 25 2006)
6. Hinsely D Answering the call, military efforts set the pace for mobile medical
    imaging July 04 www.medicalimagingmag.com (Accessed September 10 2005)
7. Teleradiology helps hospitals alleviate staff shortages www.ihealthbeat.org
    July 18 2006 (Accessed August 10 2006)
8. Marescaux J et al Transatlantic robot-assisted telesurgery . Nature
    2001;413:379-380
9. REACHMD www.reachmdc.com (Accessed January 26 2007)
10. Envision Telepharmacy www.envision-rx.com (Accessed March 3 2007)
11. Thacker PD Physician-Robot Makes the Rounds. JAMA 2005;293;150
12. Roberts R. Robots on Rounds. Kansas Business Journal Sept 5 2005 (Accessed
    September 10 2005)
13. Robotic Doctor Makes Rounds in Baltimore. www.lhealthbeat.org Feb 27 2006
    (Accessed March 10 2006)
14. Leapfrog. www.leapfroggroup.org (Accessed September 10 2005)
15. Breslow MJ Effect of a multiple-site intensive care unit telemedicine program
    on clinical and economic outcomes: An alternative paradigm for intensivist
    staffing Crit Care Med 2004;32:31-38
16. E-ICU Solution http://www.visicu.com/index_flash.asp (Accessed September
    15 2005)
17. VISICU Introduces Critical Care Without Walls. Patient Safety and Quality
    Healthcare. December 7 2006 www.psqh.com/enews/1206r.shtml (Accessed
    December 8 2006)
18. Telemedicine: A guide to assessing telecommunications in health care. Marilyn
    Field ed. National Academies Press 1996.
    http://www.nap.edu/catalog/5296.html (Accessed September 25 2006)
19. Healthcare Staffing Growth Assessment. Staffing Industry Strategic Research.
    June 2005 http://media.monster.com/a/i/intelligence/pdf(Accessed
    September 25 2006)
20. Health Buddy® www.healthhero.com (Accessed September 13 2005)
21. McGee. InformationWeek July 6 2005 (Accessed September 10 2005)
22. HomMed www.HomMed.com ihealthbeat.org April 18 2006 (June 10 2006)
23. Ross P Managing Care thru the Air. IEEE Spectrum December 2004 pp. 26-31
24. Advances in home monitoring technology Wall Street Journal Dec 12 2005
    (Accessed December 15 2005)
25. Telemanagement Center for aging services technologies www.agingtech.org
    (Accessed March 10 2006)
26. Informatics for Diabetes Education and Telemedicine
    http://www.ideatel.org/ (Accessed September 10 2005)
27. India To Launch Telemedicine Satellite www.ihealthbeat.org March 18 2005
    (Accessed September 10 2005)
28. Virginia Physicians To Set Up Telemedicine Stations in Pakistan
    www.ihealthbeat.org January 17 2006 (Accessed January 21 2006)
29. Braun PR Telemedical Wound Care Using a new generation of mobile
    telephones Arch of Derm 2005;141:254-258
30. Rural Georgia Hospitals To Receive Telemedicine Funds www.ihealthbeat.org
    December 2 2004 (Accessed September 10 2006)
31. Texas Telemedicine Used To Treat Rural Patients, Prisoners. November 30
    2006. www.ihealthbeat.org. (Accessed December 1 2006)

32. Austin M. Telehealth a virtual success. Denver Post October 24 2005 (Accessed November 10 2006)
33. Intel Announces Chinese Telemedicine Program. November 1 2006 www.ihealthbeat.org (Accessed November 2 2006)
34. Telemedicine Network Facilitates Burn Care in Three States www.ihealthbeat.org February 17 2006 (Accessed February 18 2006)
35. Psychiatrists Use Telemedicine to Treat Rural Patients www.ihealthbeat.org June 8 2006 (Accessed June 9 2006)
36. VHA Teleretinal Imaging Program www.va.gov/occ/Teleret.asp (Accessed February 7 2007)
37. Federal Communications Commission www.fcc.gov (Accessed December 20 2006)
38. USDA Rural Development http://www.usda.gov/rus/telecom/index.htm (Accessed April 12 2007)
39. Medicare payment of telemedicine and telehealth services. May 15 2006 http://www.atmeda.org/news/Medicare_Payment_Of_%20Services_Provided_V ia%20_Telecommunications.pdf (Accessed January 26 2007)

Chapter 15: Picture Archiving and Communication Systems (PACS)

Learning Objectives

After reading this chapter the reader should be able to:
- Describe the history behind digital radiology and the creation of picture archiving and communication systems
- Compare and contrast the benefits of digital radiology to clinicians, patients and hospitals
- List the challenges facing the adoption of picture archiving and communication systems
- Describe the difference between computed and digital radiology

Introduction

The following is a detailed definition of PACS:

> *"Systems that facilitate image viewing at diagnostic, reporting, consultation, and remote computer workstations, as well as archiving of pictures on magnetic or optical media using short or long-term storage devices. PACS allow communication using local or wide-area networks, public communications services, systems that include modality interfaces, and gateways to healthcare facility and departmental information systems".*[1]

Many hospitals and radiology groups are making the transition from analog to digital radiography. To their credit, radiologists have pushed for this change for years but have had to wait for better technology and financial support from their healthcare organizations. We are now at a point where the technology is mature but the financing is still an issue. Moreover, with the ongoing introduction of electronic health records (EHRs) in many health care organizations, there is a need to integrate EHRs with hospital information systems (HISs) and radiology information systems (RISs).

PACS History
- First reference to PACS occurred in 1979 when Dr Lemke in Berlin published an article describing a similar concept
- The work on the radiology data standard DICOM began in 1983 by a team lead by Dr Steven Horii at the University of Pennsylvania
- Digital imaging appeared in the early 1970's by pioneers such as Drs Sol Nudelman and Paul Capp

- University of Maryland hospital system was the first to go "film less" in 1999
- Father of PACS in this country felt to be Andre Duerinckx MD PhD [2]

PACS Basics

- Initially hospitals purchased film digitizers so routine x-ray film could be converted to the digital format
- Now digital images go from the scanning device <u>directly</u> into the PACS
- PACS usually has a central server that serves as the image repository and multiple client computers linked with a local or wide area network (LAN or WAN)
- Images are stored using the Digital Imaging and Communications in Medicine (DICOM) standard (see chapter on Interoperability)
- PACS is made possible by faster processors, higher resolution monitors, more robust hospital information systems, better servers and faster Internet connections. PACS is also aided by voice recognition to expedite report turnaround
- Input into PACS can also be from a teleradiology site via satellite
- Most monitors are still grayscale as they have better resolution (3-5 megapixels), compared to color. Newer "medical monitors" have 2,048 x 2,560 pixel resolution and can display 1000+ shades of grey instead of the 250 shades of grey seen on a standard desktop monitor
- It is estimated that about 25-30% of hospitals now have PACS [3-4]

PACS Key Components (Figure 15.1)
- **Digital acquisition devices:** the devices that are the sources of the images. Digital angiography, fluoroscopy and mammography are the newcomers to PACS
- **The Network:** ties the PACS components together
- **Database server:** high speed and robust central computer to process information
- **Archival server:** responsible for storing images. A server enables short term (fast retrieval) and long term (slower retrieval) storage
- **Radiology Information system (RIS):** system that maintains patient demographics, scheduling, billing information and interpretations
- **Workstation or soft copy display:** contains the software and hardware to access the PACS. Replaces the standard light box or view box [5]

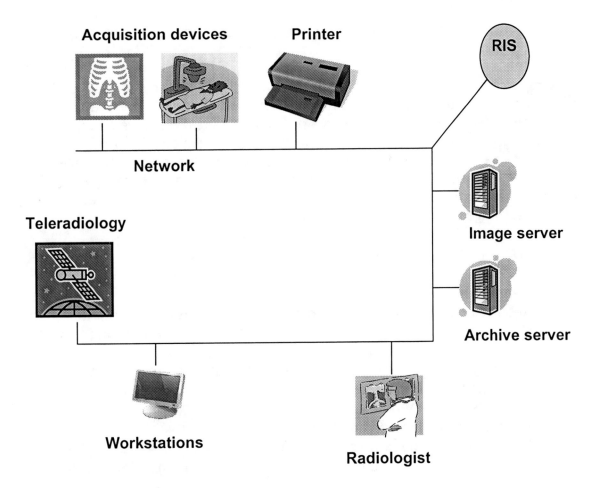

Figure 15.1. PACS Components

Types of digital detectors

1. Computed radiography: after X-ray exposure to a special cassette, a laser reader scans the image and converts it to a digital image. The image is erased on the cassette so it can be used repeatedly.[5] (Fig.15.2)

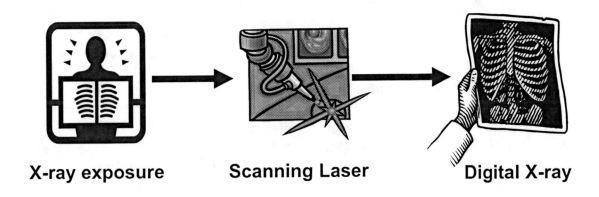

Figure 15.2. Computed Radiography

2. Digital radiography: does not require an intermediate step of laser scanning [5]

It is important to note that many facilities with digital systems or PACS still print hard copies or have some non-digital services. This could be due to physician resistance, lack of resources or the fact that it has taken longer for certain imaging services such as mammography to go digital. *Full PACS* means that images are processed from ultrasonography (US), magnetic resonance imaging (MRI), positron emission tomography (PET), computed tomography (CT) and routine radiography. *Mini-PACS*, on the other hand, is more limited and processes images from only one modality. [6]

PACS advantages and disadvantages

PACS advantages
- Replaces standard x-ray film archives which means a much smaller storage space. Space can be converted into revenue generating services and it reduces the need for file clerks
- Allows for remote viewing and reporting; to also include Teleradiology
- Can expedite the incorporation of medical images into an electronic health record system
- Images can be archived and transported on a CD-ROM
- Unlike conventional x-rays, digital films have a zoom feature
- Improves productivity by allowing multiple clinicians to view the same image from different locations
- Rapid retrieval of digital images for interpretation and comparison with previous studies
- Quicker reporting back to the requesting clinician
- Digital imaging allows for computer aided detection (CAD)
    - Using artificial intelligence, CAD identifies mammogram abnormalities
    - CAD appears to be about as accurate as the interpretation by a radiologist
    - One study confirmed that experienced radiologists used CAD <u>after</u> they reviewed the images and 50% of lesions missed without CAD were detected with CAD [7]
    - More detail about CAD available at E-medicine [8]

Several studies have shown increased efficiency after converting to an enterprise PACS. In a study by Reiner, inpatient radiology utilization increased by 82% and outpatient utilization by 21% after transition to a film less operation, due to greater efficiency. [9] In another study conducted at the University of California Davis Health System, transition to digital radiology resulted in: a decrease in the average image search time from 16 to 2 minutes (equivalent to more than $1 million annually in physician's time); a decrease in film printing by 73% and file clerk full time equivalents (FTE's) dropped by 50% (equivalent to more than $2 million savings annually). [10] The Health Alliance Plan implemented PACS at Henry Ford Health Systems in 2003. Results indicate: turn around time for films dropped from 96 hours to 36 minutes; net savings of $15 per film and key players noted significant time savings. [11]

PACS disadvantages

- Cost, cost, cost
- Integration with hospital and radiology information systems and EHR
- Bandwidth limits
- Workstation limits
- Viewing digital images a little slower than routine x-ray films
- Black and white computer monitors not as bright as traditional x-ray view boxes [7]

Conclusion

The status of PACS is similar to the issue with electronic health records. PACS is now held in fairly high regard by radiologists and healthcare organizations but CEOs and CIOs worry about the initial price tag. Unlike the EHR, there is much less of a concern about acceptance, implementation and training. The assumption is that within several years there will be a significant return on investment. PACS is an inevitable technological evolution like wireless connectivity but financial obstacles will likely delay widespread implementation in the immediate future.

References

1. Vidar corp. http://www.filmdigitizer.com/about/news/glossary.htm (Accessed April 14 2006)
2. Wiley G. The Prophet Motive: How PACS was Developed and Sold http://www.imagingeconomics.com/library/tools/printengine.asp?printArticleID=200505-01 (Accessed April 14 2006)
3. Oosterwijk HT PACS Fundamentals 2004 Aubrey Tx, Otech, Inc http://www.psqh.com/janfeb05/pacs.html (Accessed February 20 2006)
4. Gerber C. The Best on Display. Military Medical Technology Vol 11. Issue 1 2007
5. Samei, E et al Tutorial on Equipment selection: PACS Equipment overview Radiographics 2004; 24:313-34
6. Bucsko JK. Navigating Mini-PACS Options. Set sail with Confidence. Radiology Today. http://www.radiologytoday.net/archive/rt_071904p8.shtml (Accessed January 11 2007)
7. Krupinski J Digital Issues E-medicine www.emedicine.com (Accessed April 22 2006)
8. Ulissey MJ. Mammography-Computer Aided Detection. E-medicine. January 26 2005 www.emedicine.com (Accessed January 7 2007)
9. Reiner BI et al . Effect of Film less Imaging on the Utilization of Radiologic Services. Radiology 2000;215:163-167

10. Srinivasan M et al Saving Time, Improving Satisfaction: The Impact of a Digital Radiology System on Physician Workflow and System Efficiency. J Health Info Man 2006;21:123-131

11. Innovations in Health Information Technology. AHIP. November 2005. www.ahipresearch.org (Accessed January 10 2007)

 # Chapter 16: Bioinformatics

Learning Objectives

After reading this chapter the reader should be able to:
- Define Bioinformatics and how it interfaces with Medical Informatics
- State the importance of Bioinformatics in future medical treatment
- Describe the Human Genome project and its many important implications
- List the multiple private and governmental Bioinformatics databases
- Describe the application of Bioinformatics in genetic profiling of individuals and large populations

Introduction

A commonly quoted definition of Bioinformatics is:

"the field of science in which biology, computer science and information technology merge to form a single discipline" [1]

In the past two decades the field of Bioinformatics has been involved with the creation of biological databases that help evaluate DNA sequences and other genetic proteins. Scientists can study the genetic information in the databases or add new information. The process of interpreting genetic data is referred to as **computational biology** and uses algorithms and artificial intelligence to:
- Find the genes of various organisms
- Predict the structure and/or function of newly developed proteins
- Develop protein models
- Examine evolutionary relationships [2-3]

There are other bioinformatics terms worth defining:
- **Genomics:** field that analyzes genetic material from a species
- **Proteomics:** study of gene expression at the level of proteins
- **Pharmacogenomics:** study of genetic material to look for drug targets

At this time Bioinformatics seems distinct from Medical Informatics but when genetic defects are routinely diagnosed and treated that may change. The patient's genetic profile will be one more data field in the electronic health record. In late 2006 the Veterans Affairs healthcare system began collecting blood to generate genetic data that it will link to its EHR. The goal is to bank

100,000 specimens as a pilot project and link this information to new drug trials.[4] Similarly, Kaiser Permanente has created the Research Program on Genes, Environment and Health. In the first phase 2 million members will be surveyed to determine their medical history, exercise and eating habits. The second phase (2008) will require the voluntary submission of genetic material. [5] With these new genetic databases, it seems likely in the future that many treatments will be, in part, based on genetic information.

Bioinformatics Programs

The Human Genome Project (HGP)

One of the greatest accomplishments in medicine and the field of Biology is the Human Genome Project. This international collaborative project, sponsored by the US Department of Energy and the National Institutes of Health, was started in 1990 and finished in 2003. In the process of evaluating the human genome (complete set of DNA sequences for the 23 chromosomes), investigators compared their DNA results with those of the fruit fly, the bacterium E.coli and the house mouse. In addition, the HGP addressed the ethical, legal and social issues associated with the project. Now that the HGP has been completed, it will take many years to analyze and learn from the databases. [6-8] Below in figure 16.1 is the DNA sequencing of just chromosome number 12. Huge relational databases are necessary to store and retrieve this information. New technologies such as DNA arrays (gene chips) help speed the analysis and comparison of DNA fragments. [9]

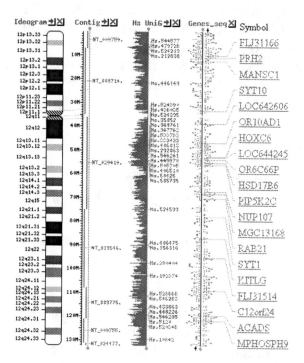

Figure 16.1. DNA sequences from chromosome 12 (courtesy National Library of Medicine)

How can Bioinformatics be useful today?

Besides diagnosing the 3000-4000 hereditary diseases that exist today, bioinformatics may be helpful in the following areas:

- Protein research to discover more targets for future drugs
- Pharmacogenomics to personalize drugs based on genetic profiles
- Complete genetic profiles will lead to better preventive medicine tests
- Gene therapy to treat diseases such as cancer. The most common way to achieve this is to use genetically altered viruses that carry human DNA. This approach, however, has not been proven to be helpful and not approved by the FDA
- Microbial genome alterations for energy production (bio-fuels), environmental cleanup, industrial processing and waste reduction
- Genetically engineered drought and disease resistant plants
- In spite of these interesting areas, it is estimated that less than 0.01% of microbes have been cultured and characterized. [5-7] As an exception, the complete genome for the common human parasite Trichomonas vaginalis was reported in the January 2007 issue of the journal Science [10]

Patients will want to know their own genetic profile even if the consequences are uncertain. Companies such as *Celera Genomics* will take advantage of the genomics project to offer genetic mapping services and pharmacogenomics.[11] *DNA Direct* is another company that offers online genetic testing and counseling. They do offer both patient and physician education and have staff genetic counselors.[12] New technologies and specialties will need to be developed and numerous ethical questions will arise. Just finding the abnormal gene is the starting point. Genetic tests will have to be highly sensitive and specific to be accepted. Patients will not be willing to undergo a preventive mastectomy or prostatectomy to prevent cancer unless the genetic testing is nearly perfect.

Decode Genetics Corporation will collect disease, genetic and genealogical data for the entire population of Iceland. Their goal is to develop better drugs based on genetic profiles to treat: heart disease, obesity, schizophrenia, pain, asthma, peripheral vascular disease and type 2 diabetes.[13] *Oracle Corporation* will partner with the government of Thailand to develop a database to store medical and genetic records. This initiative was undertaken to offer individualized "tailor made" medications and to offer bio-surveillance for future outbreaks of infectious diseases such as avian influenza.[14] Harvard University has developed a new program "Informatics for Integrating Biology at the Bedside" to analyze 2.5 million patient records to look for links between DNA and illnesses such as asthma. It is known that certain patients respond poorly to standard asthma medications and the root may be genetic. Artificial intelligence will be used to search medical records for terms such as asthma and smoking.[15]

National Center for Biotechnology Information

The NCBI was created in 1988 and is part of the National Library of Medicine and the National Institutes of Health. They host 10 different genetic databases and thereby are considered the world's largest biomedical research center. The NCBI provides access to the complete genomes of over 1,000 organisms. Genomes represent both completely sequenced organisms and those for which sequencing is still in progress. NCBI databases are listed below in Figure 16.2.

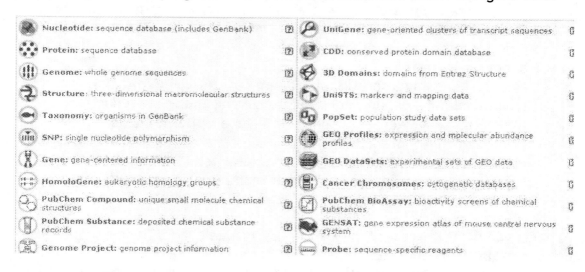

Figure 16.2. NCBI Databases (courtesy National Library of Medicine)

If you access the Genome project as listed above you can do a search for specific genes or proteins from different species. Figure 16.3 demonstrates the result of an Entrez Gene search for a tumor protein (TP53).[1]

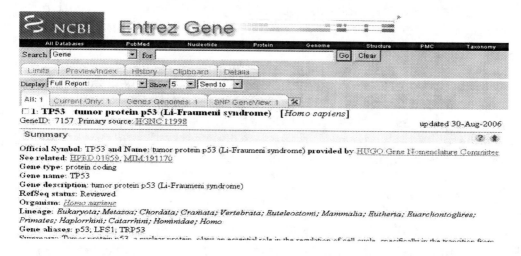

Figure 16.3. Search for a specific protein (courtesy National Library of Medicine)

The NCBI site also includes the search engine BLAST (basic local alignment search tool) that compares nucleotide or protein sequences to sequence databases and calculates the statistical significance of matches. [16]

GenBank is the NIH genetic sequence database that is a collection of all publicly available DNA sequences (100 gigabases). Interestingly, many medical journals require submission of sequences to a database prior to publication and this can be done with NCBI tools such as BankIt.[17]

Merck Gene Index

Private industry has recognized the tremendous potential of bioinformatics in research. In 1995 Merck Research Laboratory in collaboration with Washington University released to the public 15,000 gene sequences. Their ongoing releases will go to GenBank for international genetic researchers to develop future therapeutic agents. [18]

The Human Gene Mutation Database

The HGMD is a British site that attempts to collate human genetic mutations that cause inherited diseases. This has practical significance for clinicians, researchers and genetic counselors. [19]

The Online Mendelian Inheritance in Man

This is another NCBI database of genetic data and human genetic disorders. It is sponsored by Johns Hopkins University and Dr. Victor McKusick, a pioneer in genetic metabolic abnormalities. It includes an extensive reference section linked to PubMed that is continuously updated as well as a search engine. [20]

World Community Grid

Project was launched by IBM in 2004 and simply asked for people to donate idle computer time. By 2007 over 500,000 computers were involved in creating a super-computer used in Bioinformatics. Projects include the Help defeat Cancer, Fight AIDS@Home, Genome Comparison and Human Proteome Folding projects. This grid will greatly expedite biomedical research by analyzing complex databases more rapidly. [21-22]

Pharmacogenomics Knowledge Base

This organization created by Stanford University looks at the relationship between genetics, disease and drugs. There are sections on drugs, literature, variant genes, pathways, diseases and phenotypes that are searchable. [23]

For more information on Bioinformatics and genetic databases, we refer you to the classic Medical Informatics textbook by Shortliffe and Cimino.[24]

Conclusion

At this point the Human Genome Project and Bioinformatics will seem foreign to most clinicians. When they can access data that tells them who should be screened for certain cancers and which drugs are effective in which patients, these developments will be part of their day to day practices. In the meantime, scientists and biomedical companies will continue to add to the many genetic databases, develop genetic screening tools and get ready for one of the newest revolutions in medicine.

References

1. NCBI. A Science Primer. www.ncbi.nlm.nih.gov/About/primer/bioinformatics.html (Accessed July 1 2006)
2. Biotech: Bioinformatics: Introduction www.biotech.icmb.utexas.edu/pages/bioinform/Blintro.html (Accessed July 10 2006)
3. Bioinformatics Overview. Bioinformatics Web www.geocities.com/bioinformaticsweb/?200630/ (Accessed July 6 2006)
4. Ferris N. VA to launch large-scale genetic data collection. Dec. 27 2006 www.govhealthit.com (Accessed January 3 2007)
5. Kaiser Seeks Member's Genetic Info for Database. www.ihealthbeat.org February 15 2007 (Accessed February 16 2007)
6. Human Genome Project www.ornl.gov/sci/techsources/Human_Genome/project/info.shtml (Accessed July 5 2006)
7. Human Genome Project www.genome.gov (Accessed July 20 2006)
8. NCBI Human Genome Resources www.ncbi.nlm.nih.gov/genome/guide/human/ (Accessed July 19 2006)
9. DNA Arrays http://en.wikipedia.org/wiki/Dna_array (Accessed December 5 2006)
10. Carlton JM et al Draft genome sequence of the sexually transmitted pathogen Trichomonas vaginalis Science 2007;315:207-212
11. Celera Genomics http://www.celera.com/celera/about (Accessed October 2 2005)
12. DNA Direct www.dnadirect.com (Accessed November 3 2005)
13. DeCODE genetics http://www.decode.com (Accessed November 3 2005)
14. Oracle and Thai Government to build medical and genetic database www.ihealthbeat.org July 13 2005 (Accessed August 1 2005)
15. Cook G Harvard Project to scan millions of medical files The Boston Globe www.boston.com July 3 2005 (Accessed September 1 2005)
16. NCBI BLAST http://www.ncbi.nlm.nih.gov/BLAST/ (Accessed December 10 2006)
17. GenBank www.ncbi.nlm.nih.gov/Genbank/ (Accessed December 10 2006)
18. Merck Gene Index www.bio.net/bionet/mm/bionews/1995-February/001794.html (Accessed December 11 2006)
19. Human Gene Mutations Database http://www.hgmd.cf.ac.uk/docs/new_back.html (Accessed December 20 2006)
20. The Online Mendelian Inheritance in Man. http://www.ncbi.nlm.nih.gov/entrez/query.fcgi?db=OMIM (Accessed December 9 2006)
21. World Community Grid www.worldcommunitygrid.org (Accessed March 5 2007)
22. Massive computer grid expedites medical research March 14 2007 www.ihealthbeat.org (Accessed March 14 2007)
23. Pharmacogenomics Knowledge Base http://www.pharmgkb.org/ (Accessed December 20 2006)
24. Shortliffe E and Cimino J (eds) Biomedical Informatics, Computer Applications in Health Care and Bioinformatics. 3rd edition. 2006. Springer Science and Media, LLC. New York, New York.

Chapter 17: Public Health Informatics

Learning Objectives

After reading this chapter the reader should be able to:
- Define the scope and goals of public health informatics
- State the significance of the public health information network
- Identify the various disparate public health information networks
- Describe the current biosurveillance programs
- State the significance of syndromic surveillance for early detection of bioterrorism and natural epidemics

Introduction

The field of public health studies populations and not individuals. Public health tracks trends in the health of populations with the goal of preventing disease or detecting it early enough to initiate treatment. In order to study a large population you need information technology such as networks, databases and reporting software. The following is a frequently cited 1995 definition of Public Health Informatics:

> *"the systematic application of information and computer science and technology to public health practice, research and learning"* [1]

Prior to 2001, Public Health reporting consisted of hospitals and clinics sending information to local health departments, who in turn forwarded information to state health departments, who sent the final data to the Centers for Disease Control (CDC) via mail or fax. This system would not suffice for epidemics or bioterrorism. The events of September 11 2001 only heightened the concern for the means to detect illnesses and perform bio-surveillance more rapidly and accurately. Paper based reporting is simply inadequate to detect subtleties in symptoms and inadequate to report large volumes of data to a central data repository. With an electronic system, artificial intelligence could detect trends and alert officials. If the United States had a national health information network linked to EHRs in every medical facility, the reporting of data would be uniform, rapid and easy to analyze. In 2006 the United Kingdom began using *QFlu* a national influenza surveillance system. The system collects data on the diagnosis and treatment of flu-like illnesses on a daily basis from over 3,000 physicians. Thanks to an almost universal health record, electronic reporting in the UK has been greatly facilitated. [2] The United States, on the other hand, has spent billions of dollars on biosurveillance and has little to show for it, largely because of disparate systems and the lack of a national health information network. [3]

Public Health Informatics has several aspects in common with the National Health Informatics Network. Both require data standards, databases, networks and decision support. The sources of public health data are also very disparate, deriving from hospitals, clinics, public health offices, labs, environmental agencies and poison control centers.[4] Similarly, they both face budgetary hurdles due to their complexity and difficulty in implementation.

The Public Health Information Network (PHIN)

The PHIN is a relatively new CDC concept with its roots starting in 2004. The goal is to link together the players involved with US Public Health, using well established data standards. The vision is to improve disease surveillance, health status indicators, data analysis, monitoring, intervention, prevention, decision support, knowledge management, alerting and the official public health response. Ideally, the PHIN would complement the NHIN. [5-6] Its first program will be the PHIN Preparedness Initiative. This initiative was funded for $849 million dollars in 2004 to improve preparedness in all states and US territories. It is estimated that about 25% will go to information technology. At this point the initiative will:

- Define the functional requirements including early event detection, outbreak management and connection of lab systems
- Identify and use industry wide data standards like HL7 and LOINC
- Make software solutions available for public health partners [7]

The National Electronic Disease Surveillance System (NEDSS) is a major component of the PHIN that will create an interoperable surveillance system between federal, state and local networks. The system will replace several older systems and gather as well as analyze data.[8] The CDC has also created an Outbreak Management System (OMS) program that is web based and might be used in the field on a lap top computer during an actual outbreak.[9] Additional grants of $15 million will come from the Robert Wood Johnson Foundation and the program office will be the Public Health Informatics Institute.[10] The CDC has also created the Health Alert Network (HAN) that will function as PHIN's health alert component. The intent is to disseminate alerts and advisories via the Internet at the state and local levels. By late 2006 all 50 states were connected to the HAN and are funded for continuation of the initiative.[11]

Epi-X is a highly secure communications network to tie together select Public Health officials (about 4,200 users) from around the United States. The system allows for rapid reporting, alerts and discussions about possible disease outbreaks. Since its inception in 2000 over 1000 disease reports have been posted to include sentinel events such as the 2002 West Nile virus outbreak. [12]

Surveillance

Biosurveillance. The CDC is not the only federal agency engaged with biosurveillance activities. The Department of Homeland Security (DHS) has established the National Biosurveillance Integration System that will combine data from the CDC, US Department of Agriculture and environmental monitoring from the program BioWatch to improve pandemic and bioterrorism detection and response. [13] BioWatch is a Homeland Security Department program that monitors bioterrorism sensors in major US cities. The sensors are co-located with EPA air quality sensors. [14]

Syndromic surveillance. An important new Public Health function is syndromic surveillance defined as "surveillance using health-related data that precede diagnosis and signal a sufficient probability of a case or an outbreak to warrant further public health response." [15] In reality, it means that symptoms are monitored (like diarrhea or cough) before an actual diagnosis is made. If, for example, multiple individuals complain of stomach symptoms over a short period of time, you can assume there is an outbreak of gastroenteritis. In addition to the obvious sources of health data, public health officials can also monitor and analyze:

- Unexplained deaths
- Insurance claims
- School absenteeism
- Work absenteeism
- Over the counter medication sales
- Internet based health inquires by the public
- Animal illnesses or deaths [16]

Initially, public health officials were very interested in detecting trends or epidemics in infectious diseases such as severe acute respiratory syndrome (SARS) and avian influenza. After the terrorist attacks and anthrax outbreak in 2001, they have had to improve biosurveillance to detect bioterrorism. The objective is to "identify illness clusters early, before diagnoses are confirmed and reported to public health agencies and to mobilize a rapid response, thereby reducing morbidity and mortality". [17] The challenge is to develop elaborate systems that can sort through the information and reduce the noise to signal ratio. The syndrome categories that are most commonly monitored are:

- Botulism like-illnesses
- Febrile (fever) illnesses
- Gastrointestinal symptoms
- Hemorrhagic illnesses
- Neurological syndromes
- Rash associated illnesses
- Respiratory syndromes
- Shock or coma

Web based surveillance systems

ESSENCE. ESSENCE is the acronym for the electronic surveillance system for the early notification of community based epidemics. The system is part of the Department of Defense Global Emerging Infections System (DOD-GEIS). It began in the national capital area in 1999 and by 2001 it was in place at military treatment facilities (MTFs). The national capital area was selected due to its increased risk of bioterrorism. Over the past three years data has been collected from 121 Army, 110 Navy and 80 Air Force installations worldwide. The syndromic surveillance data comes from outpatient encounters (standardized ambulatory data record) that include patient demographics and ICD-9 diagnostic codes. The data is sent to a centralized server in Denver, Colorado. Every 8 hours data related to the syndromes described above is downloaded and graphed to compare daily trends with historical data. Unfortunately, it still takes several days for the data to arrive at the central server. In spite of this delay, there have been several instances where the surveillance network has identified early outbreaks before local authorities were aware. Newer versions have evolved due to collaborative efforts with Johns Hopkins University Applied Physics Laboratory and the Division of Preventive Medicine at the Walter Reed Army Institute of Research. ESSENCE II incorporated civilian data. ESSENCE IV is the most current version (January 2005) with the following features:
- Chemical-Biological detectors in limited distribution
- Data from civilian emergency rooms
- Prescription data
- Data from the Veterans system
- National insurance claims data
- Over the counter drug sales
- Standard reportable diseases such as TB, meningitis, etc [18]

BioSense. This is a CDC national web based program to improve disease detection, monitoring and situational awareness for healthcare organizations in the United States. The program will address identification, tracking and management of naturally occurring events as well as bioterrorism. Through the BioIntelligence Center, the CDC will assist in the analysis of almost real-time data using advanced algorithms, statisticians and epidemiologists. This program will be part of the PHIN and use data standards such as HL7, SNOMED and LOINC. [19]

GOARN. The Global Outbreak Alert and Response Network was established in 1997 and is supported by the World Health Organization (WHO). The severe acute respiratory syndrome (SARS) outbreak in 2003 was the first opportunity for the GOARN to be utilized. Features of the system include:

- Network provides an operational framework to alert the international community about outbreaks
- A technical collaboration to pool human resources for rapid identification, confirmation and response to outbreaks [20]

FluNet is another WHO initiative that is part of the Global Atlas. The goal is to collect and analyze infectious disease data from a country and global perspective. The Atlas will collect demographic and epidemiologic data so it can be used for queries, disease mapping and access to resources. [21]

eLEXNET (electronic laboratory exchange network) is a web based information network that provides real time access to food safety analysis. Site has tools to look for trends, geographical locations, etc. Network allows food safety experts to collaborate. [22]

Conclusion

The potential of a Public Health Information Network seems obvious to most healthcare workers but does not seem to be achievable in the near future. Although the PHIN is critical for biosurveillance data reporting, it would also be important for reporting routine public health information such as influenza outbreaks and immunization status. We can only hope that the Public Health Preparedness Initiative will be a significant step forward.

References

1. Friede A, Blum HL, McDonald M. Public Health Informatics: How Information Age Technology Can Strengthen Public Health Ann Rev Pub Health 1995;16:239-252
2. QFLU: new influenza monitoring in UK primary care to support pandemic influenza planning. http://riskobservatory.osha.eu.int (Accessed November 17 2006)
3. The biosurveillance money pit. Government Health IT. November 16 2006 www.governmenthealthit.com (Accessed November 20 2006)
4. Yasnoff WA et al Public Health Informatics: Improving and Transforming Public Health in the Information Age J Pub Health Man 2000;6:67-75
5. PHIN: Overview CDC. www.cdc.gov/phin/overview.html (Accessed September 21 2006)
6. Morris T. PHIN and NHIN. CDC. http://www.ncvhs.hhs.gov/060727p11.pdf (Accessed January 10 2007)
7. Loonsk JW et al The Public Health Information Network (PHIN) Preparedness Initiative JAIMA 2006;13:1-4
8. National Electronic Disease Surveillance System www.cdc.gov/nedss (Accessed September 21 2006)
9. PHIN: Automated Exchange of data www.cdc.gov/phin/architecture/standards/Data_Exchange.html (Accessed September 28 2006)
10. Public Health Informatics Institute www.phii.org (Accessed September 5 2006)
11. Health Alert Network. CDC. http://www2a.cdc.gov/han/Index.asp (Accessed November 1 2006)

12. Epi-X http://www.cdc.gov/epix/ (Accessed March 2 2007)
13. Beckner C. National Biosurveillance Integration System Moves Forward. May 12 2006. www.hlswatch.com (Accessed November 17 2006)
14. Shea DA, Lister SA The BioWatch Program: Detection of Bioterrorism. www.fas.org/sgp/crs/terror/RL32152.html (Accessed November 17 2006)
15. Centers for Disease Control and Prevention http://www.cdc.gov/EPO/dphsi/syndromic.htm (Accessed September 20 2006)
16. Henning K What is syndromic surveillance? MMWR September 24 2004 www.cdc.gov (Accessed September 18 2006)
17. Bioterrorism Preparedness and Response: Use of Information Technologies and Decision Support Systems www.ahrq.gov/clinic/epcsums/bioitsum.htm (Accessed September 21 2006)
18. ESSENCE II www.geis.fhp.osd.mil/GEIS/SurveillanceActivities/ESSENCE/ESSENCE.asp (Accessed September 8 2006)
19. BioSense www.cdc.gov/phin/component-initiatives/biosense (Accessed March 28 2007)
20. Global Outbreak Alert and Response Network www.who.int/csr/outbreaknetwork/en/ (Accessed September 8 2006)
21. FluNet http://gamapserver.who.int/GlobalAtlas/home.asp (Accessed March 2 2007)
22. eLEXNET www.elexnet.com (Accessed March 3 2007)

 # Chapter 18: E-Research

Learning Objectives

After reading this chapter the reader should be able to:
- Identify the multiple ways information technology can improve research
- State the general benefits of research automation
- Describe the benefits of electronic collaborative web sites
- Describe the specific benefits of electronic forms
- Compare and contrast the pros and cons of PDA based electronic forms

Introduction

The following is an important quote from a person who should know something about technology and automation:

> "The first rule of any technology used in a business is that automation applied to an efficient operation will magnify the efficiency. The second is that automation applied to an inefficient operation will magnify the inefficiency"
>
> Bill Gates

One of the definitions of medical informatics cited in chapter one includes medical research:

> "Medical informatics is the application of computers, communications and information technology and systems to all fields of medicine - medical care, medical education and **medical research**" [1]

With the exception of a few medical articles and one textbook, little has been written about the role of information technology in research. [2] In fact, many research labs are still paper based. Ironically, there have been multiple advances in technology that could make research automated, seamless and almost paperless.

Potential of information technology to improve research:
- Enhanced information retrieval through search engines such as Google and PubMed
- Automation of patient information
  - Online registration

- ○ Online surveys
- ○ Online recruitment of subjects
- ○ EHR recruitment of subjects [3]
- Electronic grant submission
- Data analysis with software programs such as Statistical Package for the Social Sciences (SPSS) [4] and Matlab [5]
- Software programs like LabVIEW can control medical devices, collate all data into a Microsoft Access database and display data real-time on a monitor during a study [6]
- E-collaboration web sites
- E-forms

In this chapter we will only discuss in detail e-collaboration and e-forms.

E-collaboration

Whether it is within or between organizations, communication is very important. Traditionally, people meet face to face to discuss how they might partner to write a grant or paper or analyze data. This is relatively simple if the collaborators work in the same building or organization but difficult if they work in different states. The Internet provides the network and space to allow collaboration and communication to occur asynchronously and securely.

An appropriate web site for research can be home grown as described by Marshall and Haley in a 2000 article in the Journal of the American Medical Association. They enumerated the 10 major steps to create a secure collaborative web site and estimated its cost to be $20-35,000 with annual maintenance costs of $2500. The article also pointed out that data in a digital format allows for more rapid uploading and analysis.[7] Avidan et al reported a web based platform serving 37 medical centers in 17 European countries for a study of decision making in intensive care units. The article discussed the importance of data validation or the means to alert researchers when data is missing or incorrect. They point out four ways data can be validated to include client sided or server sided validation or both. The importance of local and remote validation is stressed to prevent missing data. Their solution used both commercial off the shelf products and Java Script programming. [8] Other authors have also published their home grown web-based solutions.[9-11]

Commercial collaborative solutions have appeared on the horizon. As an example, Simplified Clinical Data Systems hosts a web based solution that includes the ability to create electronic forms. Some of the features of this research platform are as follows:
- Web site serves as a repository for all data collection tools, data storage and all documents related to the study
- Remote data entry via the Internet

- Online subject randomization
- Electronic case report forms (eCRFs) that are fee based
- Data validation and audit trail
- Electronic signatures
- Automated real-time (e-mail) notifications for enrollment, adverse events, protocol deviations, subject visits, etc.
- Integration with a wide variety of databases: Oracle, SQL Server, MS Access, etc.
- Customized reports
- 128 bit SSL encryption for all system transactions
- Collaborators must log on with 3 types of information to include a 6 digit number contained on a key fob that changes every minute [12]

A 2004 article in The Journal of Urology by Lallas outlines how this commercial web based product improved efficiency and collaboration among twenty one participating institutions. He concluded that the program became more cost effective as the number of enrolled subjects increased since eCRFs once created does not increase costs with more subjects. 83% of participants rated the new way to collaborate as excellent or satisfactory. [13]

ClickCommerce is another commercial comprehensive administrative research software program. It includes additional tools for Institutional Review Board requirements, grants and contracts management, e-forms library and several other research functions. [14]

A commercial off the shelf (COTS) e-collaboration solution exists with Microsoft SharePoint. This software program is part of Microsoft's 2003 Server. SharePoint Server 2007 should be released officially in early 2007. SharePoint is aimed at business collaboration and not specifically research. It creates a "do-it-yourself" web site with multiple tools and integration with Microsoft Office systems. The program requires minimal knowledge of technology and no programming skills. Many of the important features of SharePoint 3.0 include:
- Single work space to post all documents, schedules and discussions
- Document management to include data validation, document editing, audit trails, and document specific security levels
- E-mail alerts, surveys, web logs and RSS publishing
- Synchronization offline with Microsoft Office 2007 to edit calendars, contacts, tasks and set up meetings
- Access permissions set down to the document level
- 33 templates are available for download to handle issues such as vacation, expenses, travel requests and help desks

All documents related to a research study to include the Institutional Review Board (IRB) can be stored on this central location. Figure 18.1 shows an example of a research team web site where research Microsoft Excel spreadsheets are uploaded and stored. Notice the intuitive nature of document management.[15] A disadvantage of this program platform is that, due to security reasons, many organizations may not allow remote access to their server. For that reason, other solutions will be presented.

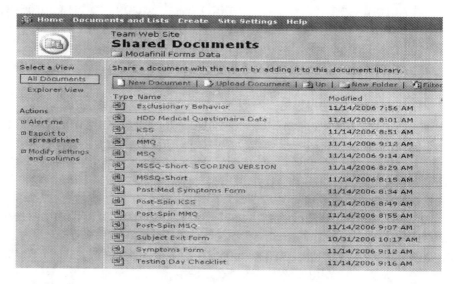

Figure 18.1. SharePoint web site for shared documents

Community Zero is a web based online community platform that could also be used for collaborative research. The program has many of the same features of Microsoft SharePoint (discussions, document uploads, group e-mails, calendars, etc.). It also requires minimal experience to establish an online community. The cost is $49.95 per year and a 30 day free trial is available.[16] OfficeZilla [17] (related to Mozilla) is a free open source web based collaborative site with many of the same features as described previously.

E-forms

Multiple problems exist with paper forms in regards to data collection, data validation, data entry and storage:
- Data collection and transcription errors may be unrecognized until after the study has been completed
- Paper forms require time to: create the forms, manually enter data and finally store data. More time required means more money spent
- Data storage takes up valuable space that could be used for more profitable ventures
- Data stored in filing cabinets limits collaboration within and outside an organization
- Data retrieval and analysis is slower with paper forms

Scannable forms. Paper forms ("bubble forms") that can be scanned by a device and tabulated have been available for some time. They are expensive to create and inflexible once published. Data is more difficult to validate and the end product is not compatible with every database.[18]

Web-based electronic forms. Forms of every description can be stored on a web site and the data automatically forwarded to a database. With data validation tools, this should result in fewer data entry and transcription errors. An alert might notify the person filling out the form that the information is incomplete so it can't be saved until corrected. Data can reside on a secure server but this still requires back up and redundancy. Web based forms have the advantage of remote input but this implies that subject testing is close to a computer. Wireless access to web based forms by a tablet PC or PDA also offers new possibilities. Additionally, patients can be recruited via e-mail and referred to forms on the web site. Complicated forms usually require programming time to add the necessary features. An example of a comprehensive e-form generating program would be OneForm Designer Plus. Customizable html forms can be created for web pages with JavaScript coding or a PDF "fillable" form. Forms can then be hosted on a web site and data automatically sent to a database.[19] Although commercial survey forms are available, so are simple free web surveys that can be used for research. [20-21]

Microsoft InfoPath is an e-forms generating software program that is part of the Microsoft Office Suite 2003 and 2007. The electronic form is usually created on a personal or laptop computer and then uploaded to a web site. InfoPath forms can be hosted on a web site, sent via e-mail or used on mobile devices. All data is written in XML (extensible markup language). Office SharePoint Server 2007 is a portal where InfoPath forms can be uploaded and managed. In that way users do not need to have InfoPath on their personal computers to complete a form. Forms creation is easy (drag-and-drop) and includes many shortcuts such as drop down menus. Forms can be created by converting Microsoft Word or Excel files. Incorrect and missing data can be detected with data validation tools. With the widespread acceptance of electronic signatures informed consent forms can now be electronic. Figure 18.2 is an example of a study form used for a drug trial. [22]

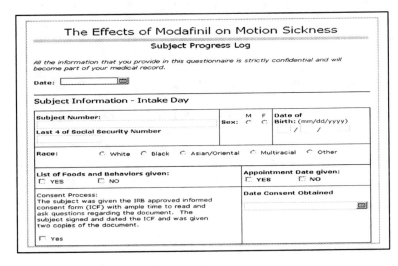

Figure 18.2. InfoPath e-form

IBM purchased PureEdge forms in 2005 and they are now known as WorkPlace forms. This comprehensive forms generating program is comparable to Microsoft InfoPath. [23]

Mobile e-forms. Handheld computers have been used for more than a decade in clinical trials.[24] Studies have suggested that this technology is accurate and fast. [25] The main advantages they offer in research are their mobile nature, low cost and small form factor. Data can be collected anywhere, including the field and later synchronized to a computer and uploaded to a database. The disadvantages are a small screen and relatively short battery life. PDA forms can be created by programming at considerable expense, but recently, commercial products exist that allow the average user to create forms. One of the products is Pendragon forms that can build a form using 23 common field types that includes images and signatures. Form creation does not require any programming experience and a two week download trial is available. The collected data inputs to an Access database on the PC, although an enterprise edition can synchronize data to a remote server.[26] Figure 18.3 shows two fields used for a study with PDA based forms. Note that with large bold font and intuitive instructions this format is suitable for simple data collection.

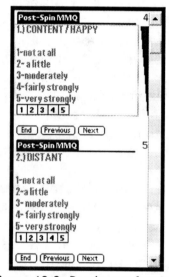

Figure 18.3. Pendragon forms

Conclusion

The evidence points towards improved productivity and accuracy using electronic collaboration tools and forms. It also seems likely that information technology can reduce full time equivalents (FTEs) due to fewer steps in data entry and data storage. The literature also suggests that data integrity is enhanced by automated data entry and validation. Multiple commercial products now exist to move research into the information age.

References

1. MF Collen, Preliminary announcement for the *Third World Conference on Medical Informatics, MEDINFO 80*, Tokyo
2. E-Research: Methods, Strategies and Issues. 2002. Anderson T and Kanuka H. Ally & Bacon Publishers
3. Embi PJ et al . Effect of a Clinical Trial Alert System on Physician Participation in Trial Recruitment. Arch Int Med 2005;165:2272-2277
4. Statistic Package for the Social Sciences www.spss.com (Accessed January 25 2007)
5. Matlab www.mathworks.com (Accessed February 20 2007)
6. LabVIEW www.ni.com (Accessed December 14 2006)

7. Marshall WW, Haley RW. Use of a Secure Internet Web Site for Collaborative Medical Research JAMA 2000;284:1843-1849

8. Avidan A, Weissman C and Sprung CL. An Internet Web Site as a Data Collection Platform for Multicenter Research. Anesth Analg 2005;100:506-11

9. Sippel H, Eich HP, Ohmann C. Data collection in multicenter clinical trials via Internet: a generic system in Java. Medinfo 1998:9:93-97

10. Mezzanaa P, Madonna Terracina FS, Valeriani M. Use of a web site in a multicenter plastic surgery trial: a new option for data acquisition. Plast Recontr Surg 2002;109:1658-61

11. Marks R, Bristol H, Conlon M, Pepine CJ. Enhancing clinical trials on the Internet: lessons from INVEST.Clin Cardiol 2001;24:V17-23

12. Simplified Clinical Data Systems. www.simplifiedclinical.com (Accessed December 10 2006)

13. Lallas CD et al . Internet Based Multi-Institutional Clinical Research: A Convenient and Secure Option. J Urol 2004;171:1880-1885

14. ClickCommerce. www.clickcommerce.com (Accessed December 14 2006)

15. Microsoft SharePoint. http://www.microsoft.com/technet/windowsserver/sharepoint/default.mspx (Accessed December 12 2006)

16. CommunityZero. www.communityzero.com/information/ (Accessed Jan 2 2007)

17. OfficeZilla www.officezilla.com (Accessed January 20 2007)

18. Shapiro JS et al. Automating Research Data Collection. Acad Emerg Med 2004;11:1223-1228

19. Amgraf Software Technology. www.amgraf.com (Accessed December 13 2006)

20. Survey Assistant Tool. www.surveyassistant.com (Accessed December 13 2006)

21. Birnbaum. Programs to Make HTML forms for Research on the Web. http://psych.fullerton.edu/mbirnbaum/programs/ (Accessed December 13 2006)

22. Microsoft InfoPath. www.microsoft.com (Accessed December 12 2006)

23. IBM Workplace forms http://www-128.ibm.com/developerworks/workplace/products/forms/ (Accessed February 20 2007)

24. Koop A, Mosges R. The use of handheld computers in clinical trials. Control Clin Trials 2002;23:469-80

25. Lal SO et al Palm computer demonstrates a fast and accurate means of burn data collection. J Burn Care Rehab 2000;21:559-61

26. Pendragon Forms www.pendragon-software.com (Accessed December 10 2006)

Chapter 19: Emerging Trends

Learning Objectives

After reading this chapter the reader should be able to:
- Identify the features of successful technology innovations
- Describe some of the future prediction by national experts
- State the significance of increased artificial intelligence in medicine
- Identify the features likely to be found in future electronic health records
- List the innovations found at the 100 most wired hospitals that will likely permeate the healthcare system in the future

Introduction

The following quote will seem comical but in reality it reflects the evolution of technology in general.

"Computers of the future may weigh no more than 1.5 tons"
Popular Mechanics 1949

Trying to predict which technology trends will succeed or fail is difficult at best and borders on pure speculation. In this chapter we will discuss some of the more significant emerging trends in information technology and Medical Informatics. As previously noted, technology continues to evolve at a rate faster than our ability to digest and assimilate it into healthcare. What determines the long term success or failure of a technology trend is often unclear but seems to be partly related to the features listed in Table 19.1.

Features	Examples
Unique concept	VoIP, voice recognition, wireless connectivity
Saves time or money	Voice recognition, VoIP
National mandate	E-prescribing capability for Medicare patients
Affordable	Wireless capability, cell phones, USB memory devices
Convenient form factor	PDAs, cell phones, USB memory devices
Ease of operation	USB memory devices, digital media players

Table 19.1. Features and examples of technology trends

Medical Informatics is heavily influenced by the need to solve problems in the field of medicine, but it is also influenced by what new technologies are available. For instance, picture archiving and communication systems (PACS) are possible today solely because of innovations in monitors, servers, digital images and processors. Another example would be the increasing capabilities of the Internet made possible by greater bandwidth. Table 19.2 demonstrates how far we have come with just personal computer technology.[1] Integrating the latest developments in technology into the field of medicine will require that more healthcare workers become formally trained in technology. Healthcare administrators and chief information officers (CIOs) will also need additional education to understand new systems such as EHRs, PACS and RHIOs, prior to implementation.

Component	1999	2005
RAM	64 MB	1 GB
Processor speed (instructions per second)	400 million	7 billion
Circuit density (total transistors)	7.5 million	125 million
Hard Drive (gigabytes)	8 gigabytes	135 gigabyte
Internet speed (bits per second)	56,000	1 million

Table 19.2. Computer components in 1999 compared to 2005

What do the experts predict?

Health and Healthcare 2010 Study by the Institute for the Future
In this book future predictions are listed as "stormy", "long and winding" or "sunny". Some of the predicted information technology advances are as follows:

- Faster microprocessor speed
- Better data storage, warehousing and mining
- More wireless applications
- Better bandwidth
- More use of artificial Intelligence
- Better encryption
- Internet 2
- Smaller, more accurate and less expensive sensors
- Drugs designed by computers
- Home telemonitoring
- Pharmacogenomics
- Improved imaging
  - Mini-MRIs with much smaller magnets
  - Electron beam CT instead of X-ray CT
  - More 3-D images
  - Better resolution, contrast and display
  - PET scans that use tumor specific markers instead of glucose making them more accurate

    ○ Computer aided diagnostic interpretations [2]

Health Care in the 21st Century

In this interesting 2005 article by Senator Bill Frist, published in the New England Journal of Medicine, changes in medical care we expect to see by the year 2015 were presented:

- Highly sophisticated EHRs that are integrated with patient portals and national information networks
- Combination medications that you only have to take daily, weekly or monthly
- Implantable computer chips that monitor vital signs and blood chemistries
- Injectable nanorobots that correct problems such as blood vessel blockages
- Automatic transmission of hospital data to insurance company so bills are paid before the patient leaves the hospital [3]

In this chapter we will take a look at some of the technological trends in medicine that have appeared on the horizon. We should point out that this is only a partial enumeration of emerging trends.

Emerging Trends

Nanotechnology

Definition: "The creation of functional materials, devices and systems through control of matter at the scale of 1 to 100 nanometers (nanometer = one billionth of a meter) and the exploitation of novel properties and phenomena at the same scale." [4] The term was first used by Taniguichi in 1974 and actually refers to multiple different technologies. Nanotechnology is already making clothes water and stain resistant and glasses scratch resistant. [5] Nanotechnology applications in the field of medicine (Nanomedicine) include diagnostic devices, contrast agents, analytical tools, therapy, and drug-delivery vehicles. Nanomedicine will also involve the use of nanorobots that are injected into the blood stream or ingested to travel to distant sites to treat disease. Imagine combating a serious infection or removing dangerous plaque from a coronary artery using nanorobots. [6] Everything will be phenomenally small except for the bill!

Artificial Intelligence in Medicine (AIM)

AIM has gone from a vague concept thirty years ago to the development of neural networks to aid in medical diagnosis and treatment. Swartz in 1970 was very optimistic in thinking that computers would radically change the delivery of healthcare by the year 2000. He stated "It seems probable that in the not too distant future the physician and the computer will engage in frequent dialogue, the computer continuously taking note of history, physical findings, laboratory and the like....".[7] Over the past thirty years universities have used artificial intelligence to develop software programs to assist medical care or what is now referred to as clinical decision support.

Neural networks, a form of artificial intelligence, are data modeling tools that capture complex relationships and learn over time. An example of a neural network would be optical character recognition (OCR). A document is scanned and the image is converted to a format such as a Word document. Each group of pixels scanned produces a value that the OCR software recognizes and converts to text. Other examples of neural networks include: target recognition, computer aided diagnosis, voice recognition and financial forecasting. [8-12] The majority of patient information is stored in the form of free text (to include voice recognition). This makes it difficult to extract information for coding and data mining. One company, Language and Computing is using natural language processing (NLP) and natural language understanding (NLU) to extract meaningful data from free text. They have entered into an agreement with Kaiser Permanente's Southern California region to develop a solution for the evaluation and management coding process. Their product will be commercially available for primary care services March 2007.[13]

Completely Integrated Electronic Health Record (EHR)
In the future EHRs will offer multiple integrated functions. Voice recognition will be associated with macros and natural language processing so when you dictate "diagnosis: type 2 diabetes" the program will know where to place the text in the record and will associate terms with ICD 9 or SNOMED CT codes. In addition, all educational content and guidelines will be embedded within the EHR. Moreover, because the age, gender, race, weight, kidney and liver function and genetic profile will be part of the EHR, you will be provided with guidance about what drug is the correct choice for any given patient or disease. Home monitoring data will be automatically integrated into the patient's EHR and neural networks will monitor and interpret the results, notifying you when a patient is deteriorating at home. EHR data mining will be a gold mine for research and development of new drugs and important quality studies. Optimistically, billing and coding will be automated so there is almost simultaneous claims submission and payment. Eventually, all electronic health records and healthcare information systems will be interoperable because they are connected to the National Healthcare Information Network.

Better Imaging
All imaging devices have improved greatly in the past decade. In spite of the fact that imaging has not become less expensive over time, devices are smaller, faster and with better resolution. For example, Siemen's Somatom Sensation™ 64 CT scanner is the fastest scanner of its generation, circling the patient in 1/3 second and producing 64 images per rotation with a resolution of

0.4 millimeters. [14] One can assume that all imaging will continue to improve in the future: ultrasounds, mammograms, CT scans, MRIs, etc.

Hospitals of the future

As a rule, the more affluent hospitals are able to purchase state-of-the-art technology as a marketing edge over the competition. Every year a survey is conducted in order to nominate the United State's 100 Most Wired hospitals. Current achievements by these cutting edge hospitals will likely predict future trends in health information technology for the average hospital:

- 90% provide access to the EHR online
- 69% offer online access to nurses notes
- 88% offer lab results online
- 90% offer radiology reports online
- Most have physicians (60%) and nurses (95%) involved in IT planning and training
- Roughly 60% offer IT training as CME
- Most offer multiple patient services online such as pre-registration and disease specific self triage
- More than 75% offer wireless access to clinical information
- Much higher adoption of bar coding and RFID [15]

Perhaps one of the best known of the "most wired" hospitals is the Indiana Heart Hospital built in 2002. They spent $25 million on technology, (total budget of $65 million), yet they expect a return on investment in only 6.6 years. A portion of their estimate, however, presumes decreased medical errors and legal costs as a result of going digital. One area of cost savings is from the installation of IP telephony (VoIP). This was less expensive than installing a formal private branch exchange (PBX) system. The hospital selected GE Medical Solutions throughout, to include Centricity as their EHR solution. Due to the completely digital and wireless environment, there are no nursing stations, as nurses are located primarily in patient rooms where they have a computer for all nursing functions.[16]

Voice over Wireless Fidelity (WiFi)

There are many applications for voice over Internet Protocol including communication within a hospital over wireless area networks. The goal is to decrease the dependence on phone lines and pagers. Vocera offers a hands-free device and Avaya and Cisco offer typical handsets. Vocera is a wireless device worn around the neck that follows commands thru voice recognition. The system operates on a wireless 802.11b network and the badge device can also send and receive telephone calls. Other options include call waiting and forwarding, call recognition by name, function or group membership. The system can also include a nurse call button integration feature so a patient can communicate with their nurse.[17]

Voice recognition

As already noted, voice recognition (VR) is a form of artificial intelligence that is catching on in healthcare. The accuracy is said to be currently in the range of 98% (out of the box by an experienced VR dictator). No longer does a clinician need to practice for an extended period of time for the VR software to recognize the speech pattern. The speed at which you can dictate has also improved dramatically and is in the 100+ words per minute range. Therefore, the learning curve is no longer steep. A separate medical vocabulary program must be purchased in addition to the basic VR software. The technology has evolved to include natural language processing, templates and macros to make the process much more robust and user friendly. Most clinicians dictate into their PC with the obvious disadvantage of not being able to use the program on other PCs. This can be overcome by hosting the software on a central server or using a portable tape recorder and later syncing with the PC or using a wireless (Bluetooth) headset. There are currently only three VR vendors that are major players: IBM, Dragon and Phillips. The most recent Dragon Naturally Speaking medical version 9.0 has 14 medical subspecialty vocabularies and can network with Citrix thin clients.[18-19]

Digital pens

Inputting data "on the fly" continues to be a challenge due to workflow demands and the need to offer more than one option to resistant physicians. One low cost option might be the use of digital pens. Logitech developed a pen that has a camera to capture data on special digital paper (expensive). It holds 2 megabytes or 40 pages of information. The pen is then synchronized to the PC. In a study by Anthem Blue Cross and Accenture, they were able to demonstrate a reduction in claims turn around time from 4-6 weeks to 2 weeks by this method. In this study, paper claims cost $2.50 per transaction compared to 49 cents using the digital pen. In a second phase they will test a wireless solution that uses cell phones to transmit the data to the insurer.[20-21] Digital pads are also an interesting alternative to routine handwriting. These clip board like devices use routine paper and digital pens to capture handwriting and images and convert them to text on the PC. Instead of the $2,000 price tag for a tablet PC, most sell for about $150-200.[22]

Smartcards

France determined that it needed a universal identification card for healthcare. In 1997 they developed a smartcard that has been deployed to over 57 million consumers (Figure 19.1). A different card is also given to all healthcare professionals to allow access to a patient data warehouse. As a result of the smartcard, reimbursement has been shortened from 6 weeks to 2-3 days. Spain, Germany, Czech Republic and Russia are also experimenting with this type of health transaction card. Each card has a microprocessor that allows them to store, process and exchange data. They carry enough information

(32K) for an emergency and unlike most credit cards, they can be read without contact. [23-24]

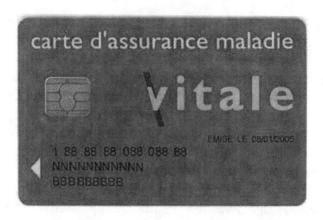

Figure 19.1. Vitale smart card

Medical smart cards have been slow to enter the American healthcare scene. Traditionally they have been used for identification and authentication. With improved memory and other features we can expect them to find new niches. [25] Florida eLife Card offers 4 MB of memory with the ability to store a complete medical history to include living wills. [25] If smart cards can be shown to expedite insurance claims payments then adoption can't be too far behind.

USB Memory Devices
Memory in all formats continues to improve. Samsung has released a 16 gigabyte compact flash device that could compete with mini-hard drives. [27] USB drive memory likewise has expanded to the 1-4 gigabyte range. It is estimated that over 130 million USB flash drives will be sold in 2007. [28]

Radio Frequency Identification (RFID)
RFID continues to grow in popularity, particularly as prices drop. The RFID industry is predicted to grow to $ 5.9 billion by 2009. RFID *passive* (no battery) tags are now much cheaper ($.25-.50). [29] In the field of medicine they are primarily used for tracking of patients, medications and assets. A new twist is that *active* RFID devices can be connected to Zigbee networks. These low cost wireless personal area networks (WPANs) transmit each others messages, thus bypassing the wired network. [30] Although *active* RFID tags are not cheap, they are durable with long battery life. Many authorities believe RFID tags will slowly replace bar coding for most common applications. Passive RFID tags will undoubtedly have more memory in the future and therefore find new indications for use. RFID receivers can be linked to existing WiFi networks which reduces cost. Hospitals have begun to use active RFID tags to track staff, patients and assets. [31] It is anticipated that in the future RFID tags will integrate sensors capable of transmitting temperature, decibels, etc. The future of injectable RFID devices for patient tracking is less certain. Although this strategy has become a standard way to identify pets, it is in its infancy

with humans. Horizon Blue Cross and Blue Shield of New Jersey plans a pilot study to implant VeriChips into patients with chronic diseases to access medical information. The chip stores an identification number that correlates with an online patient medical information database. They are in the process of seeking 280 volunteers for the program.[32]

Cell phones

It seems predictable that future cell phones will be so robust that they will become a major means of pushing and pulling information, in addition to telephonic communication. In the hospital setting they have the potential to replace pagers and access patient information from the EHR or central data repository and view patient monitoring real time. They will likely connect with the computer at home and at work and be capable of paying bills. Security issues, however, will remain a rate limiting factor. Another area that cell phones will assist in is remote disease reporting. Mobile phones have been used to track diseases in remote areas like Rwanda. Data can be sent to a central database via cell phone, PDA or the web. [33] As a result of this successful trial and the fact that 60% of Africa has cell phone coverage, a new "Phones-for-Health" project will begin in 2007 to fight the spread of AIDS. [34]

Internet2

Internet2 is a consortium (not a network) of more than 200 universities government agencies, researchers and business groups developing applications and a network for the future. The current network is known as *Abilene* and it operates at 10 gigabits per second (100-1000 times faster than Internet 1). It is closely aligned with the National LambdaRail (NLR) connecting universities across the nation thru fiber optics. Members benefit from using the faster Internet to communicate and from the development of interesting middleware. Research is underway to develop programs to support digital video, authentication and security. Although there are no plans to make Internet2 available to consumers in the immediate future, its potential in the field of medicine is tremendous.[35]

Web 2.0

Although there is no strict definition of Web 2.0, most people would say it is the new use of the Internet for collaborative purposes. Rather than one person accessing one web site, it is multiple individuals taking advantage of multiple web tools and web pages. Examples would be Wikipedia, Flickr and Clinical Informatics Wiki. Collaborative activity is often associated with open source software that creates a free infrastructure in which to work. Examples would be Linux, Apache and Mozilla. Adding to this new movement would be the appearance of blogs, podcasts, vodcasts and RSS feeds. [36-40]

Clearly, Web 2.0 is beginning to affect the field of medicine. The following are examples of medical programs that are taking advantage of new Internet applications and philosophies:

- Sermo is a free collaborative site for physicians where they can post and discuss issues and problems facing medicine [41]
- Approximately 9 million adults reported reading health related blogs online in 2006 according to Manhattan Research. This trend is so new that its future direction and impact are unclear [42]
- Medical podcasts are becoming a new trend for continuing medical education and "just in time" training. The Society of Critical Care Medicine has taken the lead in this area [43]
- Massachusetts Institute of Technology has developed an "opencourseware" concept that results in posting all of its course materials free to educators and learners worldwide [44]
- The Saphire Project is pushing the envelope one step further. They are part of the Semantic Web that enables applications to search for information based on its meaning, rather than tags. The European Union is sponsoring the Saphire Project that will aid in decision making by monitoring data from wireless personal medical sensors and hospital information systems. Semantic searching overcomes the problem of interoperability [45]

Biometrics

Security continues to be an ongoing issue with all technologies that store or communicate personal health information (PHI). Several computers now offer finger print identification in lieu of signing in with a password.[46] Although this a reasonable solution for many people, certain conditions make finger prints unreliable. There is however a new technology that reads the skin and blood flow properties in the finger using semi-conductor technology. This paper thin device is inexpensive, can be added to smart cards or cell phones and available in about 2008. [47] Retinal scanning is felt to be the most accurate biometric measure but comes with a much higher price tag. Iris imaging has been shown to be highly accurate and now used for authentication in the Netherlands and United Arab Emirates since 2001. Other biometric measures being used for authentication are face and hand geometry and speech recognition.[48-49] All technologies will require high tech scanners and a subject database for verification.

Conclusion

We live in a very exciting time in terms of rapidly improving technology. It will require highly trained clinicians who are also information technology advocates to embrace and successfully implement innovations into healthcare. More research is needed to critically appraise new information technologies before they are recommended on a large scale. Better methods will be necessary to train busy clinicians and their staff. Improved productivity, patient safety and medical quality continue to be the promise. Security, high cost and human resistance, however, will continue to be the rate limiting factors for years to come in this technological balance (Figure 19.2).

Figure 19.2. Technological balance

References

1. University of Toledo http://homepages.utoledo.edu/akunnat/chap10pt2.ppt (Accessed September 10 2005)
2. Health and Healthcare 2010: Institute for the Future Wiley John & Sons Inc or http://www.iftf.org/docs/SR-794_Health_&_Health_Care_2010.pdf (Accessed November 15 2005)
3. Frist WH Shattuck Lecture: Health Care in the 21st Century NEJM 2005;352:267-272
4. Center for Nanotechnology Education and Utilization. Penn State http://www.cneu.psu.edu/edAcademicSubOV.html (Accessed September 3 2005)
5. List of Nanotechnology Applications http://en.wikipedia.org (Accessed February 23 2007)
6. Nanotechnology http://en.wikipedia.org/wiki/Nanotechnology (Accessed September 4 2005)
7. Swartz WB Medicine and the Computer: The Promise and Problems of Change NEJM 1970;1257-1264
8. Artificial Intelligence Systems in Routine Clinical Use http://www.coiera.com/ailist/list-main.html (Accessed September 10 2005)
9. Artifical Intelligence Systems in Routine Clinical Use http://www.coiera.com/ailist/list-main.html (Accessed September 3 2005)
10. What is a neural network? Neuro Solutions. http://www.nd.com (Accessed October 1 2005)
11. Szolovits P. Artificial Intelligence and Medicine. Westview Press, Boulder, Colorado 1982
12. Coiera E Guide to Health Informatics, 2nd Edition Oxford University Press 2003
13. Language and Computing www.landcglobal.com (Accessed February 22 2007)
14. Siemens http://64slice.usa.siemens.com/patients/locations.php (November 20 2005)
15. The 100 Most Wired Vroom www.hhnmag.com July 15 2004 (Accessed July 1 2005)

16. Nash K S Indiana Heart Hospital: Real Time ER www.baselinemag.com May 4 2005 (Accessed July 5 2005)
17. Vocera www.vocera.com (Accessed April 2 2006)
18. Bergeron B Voice Recognition and Medical Transcription 8/24/2004 www.medscape.com (Accessed November 25 2005)
19. Nuance www.nuance.com (Accessed August 7 2006)
20. Anthem Tests Digital Pens To Speed Claims Payments www.ihealthbeat.org July 15 2004 (Accessed August 15 2005)
21. Logitech IO Pen http://www.logitech.com/index.cfm/products/features/digitalwriting/US/EN,crid=1545 (Accessed August 15 2005)
22. Adesso Cyberpad www.adesso.com (Accessed March 20 2007)
23. French Sesam Vitale Smartcards http://www.smartcardalliance.org/pdf/about_alliance/user_profiles/French_Health_Card_Profile.pdf (Accessed May 5 2006)
24. HealthCast 2010: Smaller world, bigger expectations. PriceWaterhouseCoopers. www.pwcglobal.com (Accessed June 12 2006)
25. Smart Card Alliance www.smartcardalliance.org (Accessed August 1 2006)
26. EMIDASI www.emidasi.com
27. Singer M Flash memory closing in on hard drives? http://news.com.com/Flash+memory+closing+in+on+hard+drives/2100-1004_3-5860251.html (Accessed August 5 2006)
28. USB Flash Drive Market http://www.u3.com/platform/ (October 23 2005)
29. RFID's Second Wave www.businessweek.com August 9 2005 (Accessed August 10 2005)
30. Zigbee networks http://en.wikipedia.org/wiki/ZigBee (Accessed January 24 2007)
31. Awarix http://www.awarix.com/products.html (Accessed February 20 2007)
32. Agovino T Insurers to test implantable microchip USA Today www.usatoday.com July 16 2006 (Accessed July 20 2006)
33. Voxiva. www.voxiva.net (Accessed October 19 2006)
34. Project Taps Cell Phones To Fight AIDS in Africa February 13 2007 www.ihealthbeat.org (Accessed February 13 2007)
35. McGill MJ An Internet Upgrade www.healthcare-informatics.com Jan 12 2005 (Accessed April 20 2006)
36. Tapscott D, Williams A Wikinomics. How Mass Collaboration Changes Everything. Penguin Books. 2006. London, England
37. Web 2.0 www.wikipedia.org (Accessed January 4 2007)
38. Health Care 2.0 Government HealthIT April 2007 vol 2. No. 2. p 22-29
39. Li, Richard Open Source 101: What Does it Mean for Healthcare? www.redhat.com
40. Open Source Software: A primer for Health Care Leaders http://www.chcf.org/documents/ihealth/OpenSourcePrimer.pdf March 2006 (Accessed January 1 2007)
41. Sermo. www.sermo.com (Accessed April 1 2007)
42. Survey: Consumers Read, Post Info on Health Blogs January 10 2007 (Accessed January 10 2007)
43. Savel, RH et al. The iCritical Care Podcast: A Novel Medium for Critical Care Communication and Education. JAMIA. 2007;14:94-99
44. Opencourseware. http://ocw.mit.edu (Accessed January 6 2007)
45. Saphire Project http://www.srdc.metu.edu.tr/webpage/projects/saphire/ (Accessed April 27 2007)
46. Toshiba www.toshiba.com (Accessed August 6 2006)
47. Nanoident Biometrics http://www.nanoident.com/CustomSolutions/biometrics.php (Accessed February 7 2007)
48. Biometrics www.wikipedia.com (Accessed August 6 2006)
49. Arun Ross, Salil Prabhakar and Anil JainAn Overview of Biometrics http://biometrics.cse.msu.edu/info.html (Accessed August 6 2006)

INDEX*

A

ACP Medicine 141, 163, 236

ADEs (adverse drug events) 51, 54, 89, 301-3, 317

Adverse drug events (ADEs) 51, 54, 89, 301-3, 317

Agency for Healthcare Research and Quality (AHRQ) 26, 28, 30, 33, 81, 254, 277, 296, 298

AHIC (American Health Information Community) 27, 81-2

AHLTA (Armed Forces Health Longitudinal Technology Application) 57-8, 118

AHRQ (Agency for Healthcare Research and Quality) 26, 28, 30, 33, 81, 254, 277, 296, 298

Alerts 18, 44, 48, 50, 52, 54-5, 122, 146, 177, 214, 254, 269, 302-3, 305, 322

Allergies 43, 45, 49, 52, 54, 155, 303, 305

Ambulatory Care Quality Alliance 277

American Academy of Family Physicians 255, 277

American College
  of Chest Physicians 252, 255
  of Physicians 50, 158, 218, 225, 237, 255, 277

American Health Information Community (AHIC) 27, 81-2

American Health Information Management Association 30, 33, 117

American Medical Informatics Association 11, 30-2, 34-5, 53, 193

American Telemedicine Association 340

AMIA 11, 30, 32-3

Armed Forces Health Longitudinal Technology Application (AHLTA) 57-8, 118

Artificial intelligence 276, 354, 357, 363, 388, 391

Authentication 101, 122, 393, 395, 397

Automation 372
  of patient information 373

B

Bar coding 21, 34, 214, 299, 306-7, 390, 394

Billing 18, 48, 56-7, 63, 93, 96-7, 116, 218, 389

Bioinformatics 10, 16, 180, 182, 189, 354-6, 359-60

Biosurveillance 364-6

Bioterrorism 89, 363, 366-8

Bluetooth 212-5, 391

C

CAD (computer aided detection) 350-1

Calculators 50-1, 196-8, 200, 206, 251, 300, 306

California HealthCare Foundation 66, 268

Cardiac clearance 197, 251-2

CCHIT (Certification Commission for Healthcare Information Technology) 63, 81, 94, 325

CCR (Continuity of Care Record) 90, 121

CDC (Centers for Disease Control and Prevention) 19, 181, 363, 365, 368

CDSS (clinical decision support systems) 40, 48, 50, 52-3, 64

Cell phones 51, 214, 299, 303, 335, 338, 384, 392, 394, 396

Centers
  for Disease Control and Prevention (CDC) 19, 181, 363, 365, 368
  for Medicare and Medicaid Services (CMS) 26, 265, 270, 278, 280, 336

Centre for Evidence Based Medicine in Oxford 225

Cerner 91, 96

Certification Commission for Healthcare Information Technology (CCHIT) 63, 81, 94, 325

CHCS 57-9

Clinical
  decision support systems (CDSS) 40, 48, 50, 52-3, 64
  practice guidelines (CPGs) 14, 45, 48-9, 53, 140, 156, 227, 244-9, 251-2, 254-6, 260-1, 263, 269, 281, 285

CMS (Centers for Medicare and Medicaid Services) 26, 265, 270, 278, 280, 336

Codes 51, 58, 62, 94, 97, 205, 210

Communication 10, 17, 47-8, 93, 96, 99, 123, 218, 299, 332, 346-7, 372, 374, 390
  patient-physician 35, 107-8
  systems 18, 21, 49, 330, 345-6, 384

Computer aided detection (CAD) 350-1

Computerized Patient Record System (CPRS) 58

Computerized physician order
  entry (CPOE) 25, 28, 48, 50, 52-5, 61-2, 68-9, 290, 297, 300-3
  entry in Ambulatory Settings 317

Computers, personal 12, 15, 119, 121, 192-3, 379
